CRISPR Technology - Recent Advances

Edited by Yuan-Chuan Chen

Published in London, United Kingdom

CRISPR Technology - Recent Advances
http://dx.doi.org/10.5772/intechopen.102156
Edited by Yuan-Chuan Chen

Contributors
Masahiro Sato, Masato Ohtsuka, Emi Inada, Shingo Nakamura, Issei Saitoh, Shuji Takabayashi, Maximilian Evers, Stephan Kolkenbrock, Sönke Friedrichsen, Björn Brändl, Franz-Josef Müller, Reza Mohammadhassan, Sara Tutunchi, Negar Nasehi, Fatemeh Goudarziasl, Lena Mahya, Stephane Pelletier, Annelise Cassidy, Jack C. Yalowich, Terry S. Elton, Jessika Carvajal-Moreno, Md. Ismail Hossain, Xinyi Wang, Dalton J. Skaggs, Yuan-Chuan Chen

First published in London, United Kingdom, 2023 by IntechOpen
IntechOpen is the global imprint of INTECHOPEN LIMITED, registered in England and Wales, registration number: 11086078, 5 Princes Gate Court, London, SW7 2QJ, United Kingdom

British Library Cataloguing-in-Publication Data
A catalogue record for this book is available from the British Library

Additional hard and PDF copies can be obtained from orders@intechopen.com

CRISPR Technology - Recent Advances
Edited by Yuan-Chuan Chen
p. cm.
Print ISBN 978-1-80356-815-7
Online ISBN 978-1-80356-816-4
eBook (PDF) ISBN 978-1-80356-817-1

Meet the editor

Yuan-Chuan Chen obtained his Ph.D. in Comparative Biochemistry at the University of California, Berkeley, USA, and completed postdoctoral studies at the Taiwan Food and Drug Administration and National Applied Research Laboratories. His research specialties include pharmacy/pharmacology, biochemistry, microbiology/virology, immunology, cell/molecule biology, biotechnology/nanotechnology, cell/gene therapy, and policy/regulation. He has published more than twenty co-authored articles in peer-reviewed journals and seven book chapters on basic science, biomedicine, and related policy/regulation. Dr. Chen is currently an assistant professor at the Jenteh Junior College of Medicine, Nursing and Management, Taiwan. His studies focus on biopharmaceuticals, and he is interested in the development of agricultural/industrial products and human therapeutics using CRISPR technology.

Contents

Preface

For traditional genetic cloning, the target genes are cut at a specific site using restriction endonucleases. Scientists cannot modify genomic sites wherever they want and screening for clones of interest is usually very time-consuming. Currently, a popular gene editing approach known as CRISPR (clustered regularly interspaced short palindromic repeats) technology can be used to engineer the desired genes in vitro and in vivo efficiently and precisely, without the limitation of needing a restriction site. In addition to basic research, CRISPR technology has been applied in product manufacturing, including diagnostic tools, agricultural products, foods, industrial products, and medicinal products. Novel therapeutic strategies based on CRISPR technology have the potential to bring the hope of recovery to patients with life-threatening illnesses for which effective drugs or medical devices are not available, though some technical, safety and ethical issues need to be addressed.

CRISPR technology is being improved to become more mature, specific, efficient, and safe for applications as scientists aim to maximize benefits and minimize risks. The potential advantages of this revolutionary technology are endless. This book discusses CRISPR technology and its development, technology, challenges, and applications.

We would like to thank all the authors for their valuable contributions. We also thank the staff at IntechOpen for their assistance throughout the publication of this book.

Yuan-Chuan Chen, Ph.D.
Department of Nursing and Department of Medical Technology,
Jenteh Junior College of Medicine, Nursing and Management,
Miaoli County, Taiwan

Section 1

Introduction

Chapter 1

Introductory Chapter: CRISPR Technology

Yuan-Chuan Chen

1. Introduction

The clustered regularly interspaced short palindromic repeats (CRISPR) was originally derived from bacteria fighting against foreign genetic material, such as plasmid or viral DNA. This is an adaptive immunity generated in the bacteria infected with bacteriophage. Traditionally, bacteria will have a memory of the DNA they have invaded, and when DNA with the same sequence enters the bacteria again, an acquired immune response will be generated to break down the foreign DNA. CRISPR consists of multiple, short, and direct repeats of DNA sequences, each repeat containing a series of bases accompanied by about 30 bases called spacer DNA. These spacers are short DNA fragments from plasmids or bacteriophages. When the host encounters this particular plasmid or bacteriophage again, it will recognize the foreign DNA by complementation with CRISPR RNA (crRNA). After crRNA binds to complementary foreign DNA, the Cas9 protein (nuclease) breaks down and destroys the invading DNA or RNA. The mechanism is that single-stranded guide RNA (sgRNA) interacts with Cas9, and the combination of sRNA and Cas9 will guide the endonuclease activity to the region adjacent to the protospacer sequence (PAM). After the sgRNA recognizes a specific DNA sequence, the bound Cas9 will cut 3 nucleotides upstream of the PAM (NGG) of the positive and negative DNA strands, forming a double-stranded break with a blunt end. Because CRISPR technology is becoming more mature and stable, it has been successfully applied in genetic editing, diagnosis, and medicine for years.

2. Genetic editing

Gene cloning is to select a specific target gene to manipulate the genetic traits of an organism and then use molecular biology methods to modify the target gene. The modified genome is put into the target gene and recombined into an expression vector and finally transformed into a suitable host cell for a large number of expressions. The main production processes, include vector gene cleavage, target gene cutting, recombinant vector, transformed host, recombinant vector replication, and host cell culture, including gene transfer between the same species and different species to improve or generate new animals, plants, and microorganisms.

Traditional gene cloning methods use restriction enzymes to cut specific restriction sites in the genome. After selecting a gene, the cutting position is not accurate enough, screening is time-consuming, labor intensive, and expensive. Genetic editing of organisms based on CRISPR technology has the advantages of high efficiency, accuracy, speed, and economy, compared with traditional methods. Genetic editing has revolutionized biological research through the new ability to precisely edit the genomes of living organisms. Recently, various genetic tools have been explored for engineering simple and complicated genomes. The CRISPR/Cas9 system has widely been used in genetic editing because of its high efficiency, ease of use, convenience, and accuracy. It can be used to add desirable and remove undesirable genes simultaneously in a single event. Additionally, many newly emerging CRISPR/Cas systems, such as base editor, xCas9, Cas12a (Cpf1), and Cas13 are also considered. The scientific community has already used this technology to modify human cells, animals, plants, or microorganisms. Transgenic animals and plants or engineered microorganisms are used in basic research, such as viruses, bacteria, yeasts, protozoa, plants, mice and human cells. Many literatures related to CRISPR technology have been published in the past few years, and genetic editing is currently the most successful and extensive application [1, 2].

3. Diagnosis

The traditional methods (e.g., isolation and identification, nucleic acid and antigen detection, and specific antibody detection.) used to perform diagnosis of microbe infection in humans or contamination in food are not only time-consuming, labor-intensive, and expensive but also require sophisticated equipment and the professionals. Therefore, it is necessary to develop a new intelligent and rapid diagnostic method that does not need to rely on professional equipment and personnel. CRISPR technology has been introduced into the field of rapid nucleic acid detection for the development of new medical detection tools and reagents, bringing breakthroughs to existing testing and diagnostic technologies. For example, the DNA endonuclease-targeted CRISPR trans reporter (DETECTR) system for CRISPR/Cas12a (Cpf1) enables analysis of cells, blood, saliva, urine, and feces to detect genetic mutations, cancer, and antibiotic resistance and can be used to diagnose bacterial and viral infections. We take the detection of human papillomavirus (HPV), Zika virus, dengue virus, and SARS-CoV-2 as examples to show the potential of CRISPR technology as a diagnostic tool.

3.1 Human papillomavirus

In 2018, Jennifer Anne Doudna's team at the University of California, Berkeley, used the CRISPR/Cas12a (Cpf1) system to cut the target's double-stranded DNA (dsDNA) and found that the Cas12a nuclease would be activated and nonspecifically cleave single-stranded DNA (ssDNA), which deliver the CRISPR/Cas12a system and nonspecific ssDNA fluorescent labeling (FQ-labeled reporter) into cells [3]. Once the target dsDNA is detected, the CRISPR/Cas12a system will be activated and the fluorescent reporter gene will also be degraded to release a fluorescent signal. In a previous study, they demonstrated that CRISPR can be a tool for diagnosing viral infections in vitro. The DETECTR can rapidly and accurately detect HPV infections in patient specimens. The detection rate of HPV16 and HPV18 infection is 100% (25/25 agreement) and 92% (23/25 agreement), respectively, and HPV16 and HPV18 are known to be the most dangerous subspecies causing cervical cancer [3].

3.2 Zika virus and dengue virus

In 2018, Feng Zhang's team in the Broad Laboratory of Massachusetts Institute of Technology (MIT) combined isothermal preamplification with Cas13 nuclease to detect single-stranded RNA (ssRNA) and ssDNA tool-specific high-sensitivity enzymatic reporter unlocking (SHERLOCK) to deliver a variety of enzymes and fluorescent reporter genes to cells [4]. If the CRISPR system finds the target gene, the corresponding Cas13 will activate the cleavage enzyme and specifically cut the corresponding fluorescent reporter gene, which releases fluorescent signals. The SHERLOCK can be used to detect ssRNA viruses, such as Zika virus and dengue virus in human specimens, such as saliva. After the introduction of a variety of bacterial Cas13 nucleases from different genera, such as LwaCas13a and PsmCas13b, SHERLOCK was upgraded to SHERLOCK version 2 (SHERLOCKv2). Because different Cas 13 nucleases show different degrees of "preference" for different RNA sequences, SHERLOCKv2 sensitivity is increased by 3.5 times, compared with the original SHERLOCK [4].

3.3 COVID-19 (SARS-CoV-2)

In 2020, Chinese researchers summarized the latest progress on COVID-19 detection (SARS-CoV-2 infection) based on various CRISPRs, including CRISPR/Cas9, CRISPR/Cas12, and CRISPR/Cas13, which are being developed as a fast, accurate, and portable diagnostic method, showing the potential of CRISPR to be applied in the diagnosis of COVID-19 and other emerging infectious diseases [5]. Compared with the polymerase chain reaction (PCR) assay and DNA sequencing methods, this new method can more rapidly identify the pathogens of emerging infectious diseases and facilitate timely treatment. While ideal detection reagents are characterized by being fast, reliable, inexpensive, and convenient, these emerging diagnostics still require careful testing and clinical validation to ensure their functionality [5].

In 2020, Ackerman et al. proposed a platform called CARMEN (specific high-sensitivity enzymatic unlocking and an extension of SHERLOCK), which could detect a range of pathogen infections, including the new coronavirus that causes COVID-19 [6]. The detection mixture contains sgRNA, Cas13, and a fluorescently labeled reporter RNA. The fluorescent molecule is attached to the reporter RNA and does not emit light. SgRNA can recognize a specific target nucleic acid (DNA or RNA) sequence. If the CRISPR/Cas13 complex recognizes the target nucleic acid sequence, Cas13 will be activated to cleave the reporter RNA and generate fluorescence, thereby detecting a specific virus infection in the specimen [6].

In 2020, Mammoth Biosciences and GSK announced the development of a CRISPR-based SARS-CoV-2 detection platform DETECTR, which can rapidly (less than 40 minutes) and accurately detect SARS-CoV-2 from nasal swab RNA extracts from examinees [7]. The method was validated using reference samples and clinical samples from patients, including 36 patients with COVID-19 and 42 patients with other viral respiratory infections in the United States (US). They found that the test results had a positive predictive concordance of 95% and a negative predictive concordance of 100%, demonstrating that this is a visible and rapid detection method and has the potential to replace the most widely used quantitative **real-time** reverse transcription polymerase chain reaction assay (qRT-PCR) [7]. This detection platform is characterized by being fast, easy to operate, portable, and completely disposable. It can identify SARS-CoV-2 RNA through a simple nasal swab and does not require professionals, laboratories, and instruments. This DETECTR can obtain test results within 20 minutes and detect a variety of infectious diseases. GSK submitted the entire testing platform to the US FDA for evaluation in 2020 and promoted it as a point-of-care testing tool for hospitals and clinics. The US FDA has conducted an emergency use authorization (EUA) review of the platform. The platform is expected to provide over-the-counter tests for the general public to use at home in the near future.

4. Medicine

CRISPR can be used for drug discovery and screening successfully and significantly facilitate the pharmaceutical development. However, human therapeutics based on CRISPR for which direct disease treatment was only attempted or under clinical trial due to the concerns of safety, efficacy, and ethics, including inheritance disease, viral infections, neurodegenerative diseases, metabolic diseases, and cancer. Currently, the most perspective fields for the treatment based on CRISPR are precision medicine and gene therapy.

4.1 Precision medicine

The most sophisticated and sensitive field of CRISPR development should be therapeutic. Because human therapy is related to life, health, and human rights, it requires the highest technical level, and all critical issues must be considered. The most stringent regulatory requirements and restrictions are performed around the world. The most important application of CRISPR technology in human medicine should be precision medicine, also known as personalized treatment. This concept was first proposed by the National Research Council of the US in 2011. Human individuals will show different traits due to genetic differences, and the symptoms and severity of the disease will also vary. The same treatment methods are used for different individuals with the same disease, but they may have different therapeutic effects. Therefore, it is often necessary to have different treatment strategies depending on individual differences. In addition to the patient's description of symptoms and routine examinations (e.g., blood test, X-ray examination, and ultrasound examination) that are used in traditional methods, precision medicine also includes biomedical tests, such as genetic testing, protein testing, and metabolic testing. For precision medicine, they analyze personal data (e.g., gender, height, weight, race, past medical history, family medical history, and test results.) through the human genetic database to select the most suitable strategy, and drugs for patients to maximize the therapeutic effect and minimize side effects. CRISPR technology can accurately perform gene editing and has the potential to correctly repair gene mutations, which can be applied to the clinical treatment of diseases, thus becoming precision medicine.

4.2 Gene therapy

Although CRISPR has the potential to be directly used in human therapy, most of them are only at the stage of basic research or animal experiments, and few are actually conducted clinical trials or product launches because of lack of specificity, fear of causing mutations, and difficulty in delivery tool selection. It is complicated and there are many options for treatment. For most diseases, the direct use of CRISPR- based treatment may not be effective and even lead to aggravating the disease if you rashly skip conventional treatment methods. However, there are exceptions, especially if the main cause of diseases is due to defective genes. For congenital genetic diseases, traditional therapy usually only alleviates the symptoms but not completely cure the disease, and a complete cure requires the modification or repair of genes. In these special cases, the safety and ethics concerns of treatment based on CRISPR are less, thereby the compassionate use is possible. Therefore, many scientists are focusing on developing CRISPR as a tool for gene therapy.

In 2021, the United Kingdom and New Zealand researchers revealed the drug NTLA-2001 developed from CRISPR/Cas9 can treat the rare disease "familial amyloid polyneuropathy" [8]. This disease is mainly due to gene mutation, which accumulates misfolded transporter (transthyretin, TTR). The main function of NTLA-2001 is to reduce the concentration of TTR in serum. After completing the preclinical trials in vitro and in vivo, the team conducted a clinical phase 1 trial to evaluate the safety and efficacy of a single escalating dose of NTLA-2001 on patients, with a total of six subjects [8]. Animal studies in preclinical studies have shown that the TTR gene can be permanently knocked out after a single dose. In contrast to the results of the Phase 1 clinical trial, patients were assessed for safety within the first 28 days after the infusion of the drug, and few adverse events were found. On day 28, serum TTR protein concentrations were found to be reduced by an average of 52% from baseline in patients receiving the 0.1 mg/kg dose, compared with an 87% reduction at the 0.3 mg/kg dose. In a small group of patients with hereditary ATTR and polyneuropathy, existing research results show that NTLA-2001 effectively reduces the concentration of pathogenic TTR protein in serum by targeting TTR gene through CRISPR, while only mild adverse events occur [8].

5. Conclusion

The use of CRISPR technology for genetic editing, diagnosis, drug development, and screening is extensive and less controversial because it is not directly related to human therapeutics. CRISPR has the potential to be used for the treatment of diseases, including genetic disease, viral infections, neurodegenerative disorders, metabolic diseases, cancer and regenerative medicine; however, there are still considerable safety, efficacy, and ethics concerns for the treatment based on CRISPR. Many pharmaceutical companies and biotechnology companies have begun to produce various medical products using CRISPR technology. Only few cell therapy and gene therapy products based on CRISPR have been approved on the market, and some are still under clinical trials. Currently, the conventional treatments have been tested and proven to be effective for most diseases. With the advancement of medical technology and the development of new drugs, many diseases that were considered incurable in the past have been treated well now. Therefore, patients should use conventional treatment first and consider using CRISPR-based therapy, while the disease cannot be controlled or existing drugs cannot cure the disease. According to the current testing results, most of the treatments based on CRISPR are still in the stage of accumulating experience, and more clinical data are still needed to prove their effectiveness and safety. It belongs to medium and high-risk medical behaviors. Gene therapy, using CRISPR, is likely to be a suitable method to be tested for the treatment of inherited diseases because inherited diseases can be cured only when the defective gene is modified. Before patients receive direct treatment based on CRISPR, genetic testing should be performed to confirm suitability, consultation with professional physicians must be completed, and the evaluation of whether the study meets the regulations of compassionate use is required to ensure the welfare of patients.

Author details

Yuan-Chuan Chen[1,2]

1 Department of Medical Technology and Department of Nursing, Jenteh Junior College of Medicine, Nursing and Management, Miaoli County, Taiwan

2 Program in Comparative Biochemistry, University of California, Berkeley, CA, USA

*Address all correspondence to: yuchuan1022@gmail.com

References

[1] Manghwar H, Lindsey K, Zhang X, Jin S. CRISPR/Cas system: Recent advances and future prospects for genome editing. Trends in Plant Science. 2019;**24**(12):1102-1125

[2] Zhang C, Quan R, Wang J. Development and application of CRISPR/Cas9 technologies in genomic editing. Human Molecule Genetics. 2018;**27**(R2):79-88

[3] Chen JS, Ma E, Harrington LB, Da Costa M, Tian X, Palefsky JM, et al. CRISPR-Cas12a target binding unleashes indiscriminate single-stranded DNase activity. Science. 2018;**360**(6387):436-439

[4] Gootenberg JS, Abudayyeh OO, Kellner MJ, Joung J, Collins JJ, Zhang F. Multiplexed and portable nucleic acid detection platform with Cas13, Cas12a, and Csm6. Science. 2018;**360**(6387):439-444

[5] Xiang X, Qian K, Zhang Z, Lin F, Xie Y, Liu Y, et al. CRISPR-Cas systems based molecular diagnostic tool for infectious diseases and emerging 2019 novel coronavirus (COVID-19) pneumonia. Journal of Drug Targeting. 2020;**26**:1-5. DOI: 10.1080/1061186X.2020.1769637

[6] Ackerman CM, Myhrvold C, Thakku SG, Freije CA, Metsky HC, Yang DK, et al. Massively multiplexed nucleic acid detection with Cas13. Nature. 2020;**582**(7811):277-282

[7] Broughton JP, Deng X, Yu G, Fasching CL, Servellita V, Singh J, et al. CRISPR-Cas12-based detection of SARS-CoV-2. Nature Biotechnology. 2020;**38**(7):870-874

[8] Julian D, Gillmore JD, et al. CRISPR-Cas9 In vivo gene editing for transthyretin amyloidosis. The New England Journal of Medicine. Aug 2021;**385**(6):493-502. DOI: 10.1056/NEJMoa2107454

Chapter 2

Emerging CRISPR Technologies

Annelise Cassidy and Stephane Pelletier

Abstract

The discovery and implementation of Clustered Regularly Interspaced Short Palindromic Repeats (CRISPR) and CRISPR associated (Cas) systems for genome editing has revolutionized biomedical research and holds great promise for the treatment of human genetic disorders. In addition to the popular CRISPR-Cas9 and CRISPR-Cpf1 systems for genome editing, several additional Class I and Class 2 CRISPR-Cas effectors have been identified and adapted for genome editing and transcriptome modulation. Here we discuss current and emerging CRISPR-based technologies such as Cascade-Cas3, CRISPR-associated transposases (CAST), CRISPR-Cas7–11, and CRISPR-Cas13 for genome and transcriptome modification. These technologies allow for the removal or insertion of large DNA elements, the modulation of gene expression at the transcriptional level, and the editing of RNA transcripts, expanding the capabilities of current technologies.

Keywords: CRISPR, Cas9, Cpf1, Cascade, Cas3, Cas12k, Cas13, Cas7–11

1. Introduction

Since the discovery of the double helix, scientists have been searching for ways to manipulate genomes. Over the past 15 years, technological advances such as the development of targetable nucleases finally provided a means for introducing specific alterations within a genome of interest. Targetable nucleases function by introducing a DNA double strand break (DSB) at a precise location within a genome which in turn activates cellular DNA repair pathways. By hijacking these pathways, via the coadministration of DNA repair templates, a plethora of genetic modifications ranging from single nucleotide substitutions to chromosomal translocations can be engineered (**Figure 1**).

The first implementations of targetable nucleases included zinc finger nucleases (ZFNs) and transcription activator-like effector (TALE) nucleases (TALENs). These enzymes are formed by the combination of a non-specific DNA endonuclease called FokI and DNA binding protein domains derived from the zinc finger or TALE family of transcription factors. These enzymes function as obligate dimers and rely on protein-DNA pairing for target recognition [1, 2]. While these enzymes provide the specificity needed for engineering precise DNA alterations, their programming or reprogramming necessitates the design and synthesis of a new pair of enzymes for each new alteration. The adaptation of the Class 2 type II CRISPR-Cas9 system from *Streptococcus pyogenes* (CRISPR-SpCas9) for genome editing has drastically changed the way genetic engineering is performed in that it provides the long-awaited

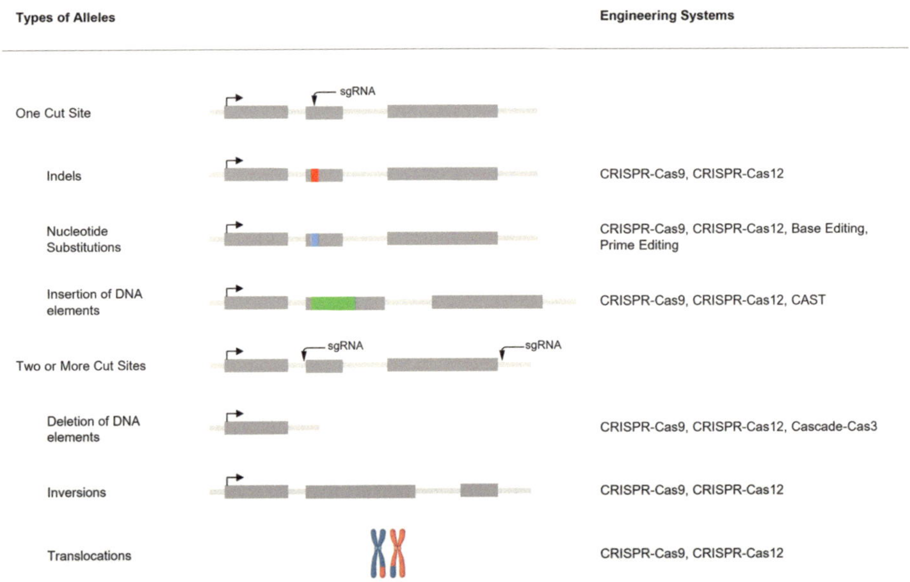

Figure 1.
Genetic modifications engineered using conventional CRISPR-Cas systems. Schematic representation of the different types of alleles that can be engineered with a single or a pair of cut sites and the engineering systems used to create them. With one cut site, three types of alleles can be engineered: Targeted random insertion or deletion of genetic material (indels), nucleotide substitutions, and insertion of DNA elements. With two cut sites, three other types of alleles can be engineered: Deletion of DNA elements, inversions, and chromosomal translocations. Introns are shown as light gray boxes, exons as dark gray boxes, start sites as black arrows. Indels are shown as a red box, nucleotide substitutions as a blue box, and insertion of DNA elements as a green box. Black arrows pointing to a specific location within an intron or exon indicate cut sites.

simplicity and versatility required for engineering precise DNA alterations. Rather than relying on protein-DNA interactions for target recognition, CRISPR-SpCas9 relies on RNA–DNA base pairing. The simple modification of a short RNA transcript is sufficient to reprogram a nonspecific endonuclease to target other sites.

In bacteria and archaea, CRISPR-Cas systems are RNA-based immune systems that control virus and plasmid invasion [3]. CRISPR-Cas systems are taxonomically classified as Class 1 and Class 2 systems based on the number of components involved in the interference stage of the immune response. With rare exceptions, Class 1 systems, which account for approximately 90% of all CRISPR-Cas systems in prokaryotes, use multiprotein effector complexes whereas Class 2 systems use a single effector. Class 1 and Class 2 systems are further divided, based on signature genes and distinctive gene architectures, into three or more types: Type I, III and IV for Class 1 systems and type II, V and VI for Class 2 systems.

Concurrent with the implementation of CRISPR-Cas9 for genome editing, a variety of Class 1 and Class 2 systems with complementary properties to type II effector Cas9 have been identified and adapted for genome and transcriptome editing. These systems include: Class 1 type I Cascade-Cas3 systems [4–6], which are RNA-guided DNA shredding systems; Class 2 type V-K effector Cas12k-Tn7-like transposase systems [7], which are RNA-guided DNA transposition systems that allow for unidirectional insertion of large DNA cargos; and Class 2 type VI CRISPR-Cas13 [8, 9] and Class 1 Type III-E effector Cas7-11 systems [10, 11], which are RNA-guided RNA targeting systems.

In this chapter, we describe conventional as well as emerging CRISPR-Cas-based technologies for transcriptome and genome editing. We provide a simplified view of these systems and their operons. When applicable, we describe the modifications made for genome editing in mammalian cells, the sequence and structure of guide RNAs, PAM requirements, and examples of their use for genome and transcriptome editing. For simplicity, we focus on CRISPR-Cas9 from *S. pyogenes* [12–14], CRISPR-Cpf1 from *Francisella novicida* [15], Cascade-Cas3 from *Thermobifida fusca* [5], CAST from *Scytonema hofmannii* [7, 16], CRISPR-Cas13 from *Leptotrichia shahii* [9], and Cas7–11 from *Candidatus Scalindua brodae* [17] as prototypic systems.

2. Conventional CRISPR-Cas systems for genome editing

Genome editing using conventional CRISPR-Cas systems functions by introducing DNA DSBs at a precise location within a target genome. These breaks, known to be highly recombinogenic, are typically repaired via the nonhomologous end joining (NHEJ) DNA DSB repair pathway and result in the random insertion or deletion of genetic material, often referred to as indels. In actively dividing cells, homologous recombination (HR) can also occur and uses the sister chromatid as repair template, resulting in error free repair of the break. By providing an exogenous repair template, in the form of single or double stranded DNA molecules, a variety of genetic alterations can be engineered, including nucleotide substitutions, insertion of DNA elements, deletion of DNA material, inversion of DNA elements, as well as chromosomal translocations (**Figure 1**). Two main CRISPR-Cas systems from various species, CRISPR-Cas9 and CRISPR-Cpf1, have been adapted for genome editing and other applications.

2.1 CRISPR-Cas9

The most frequently used CRISPR-Cas system for genome editing in mammalian cells has been derived from the Class 2 type II CRISPR-Cas system from *S. pyogenes* (**Figure 2A**) [12–14]. In this system, a large endonuclease named SpCas9 pairs with a single guide RNA (sgRNA) to target the ribonucleoprotein complex to a specific location within a genome via the formation of an RNA–DNA duplex according to Watson-Crick base pairing (**Figure 2B**). sgRNAs are 96 nucleotide-long RNA transcripts formed from the fusion of a CRISPR-RNA (crRNA), which provides target specificity, and the trans-acting crRNA (tracrRNA), which bridges the crRNA to the endonuclease SpCas9. The first 20 nucleotides of a sgRNA provides target specificity whereas the other 76 nucleotides contain sequences encoding the tracrRNA (**Figure 2C**). Binding of SpCas9 to its target site results in the formation of an R-loop, a three-stranded nucleic acid structure composed of a DNA–RNA hybrid and unbound DNA strand referred to as the protospacer element or target site (**Figure 2B**). Following the R-loop formation, Cas9 – which possesses a RuvC-like and an HNH-like domain – cleaves both DNA strands 3 nucleotides upstream of the protospacer adjacent motif (PAM). PAM sequences are short genomic sequences located at the 3′ end of the protospacer element (for Cas9 nucleases) that must be present for the nuclease to be catalytically active. Although this requirement was initially viewed as a major limitation to the application of CRISPR-SpCas9 for genome editing, its PAM sequence, 5′-NGG-3′, is found approximately every 8 nucleotides in the human genome. Moreover, several CRISPR-Cas9 systems from various bacterial species have been identified, each having distinct PAM requirements (**Table 1**) [12], expanding the scope of these systems. For genome editing in mammalian cells,

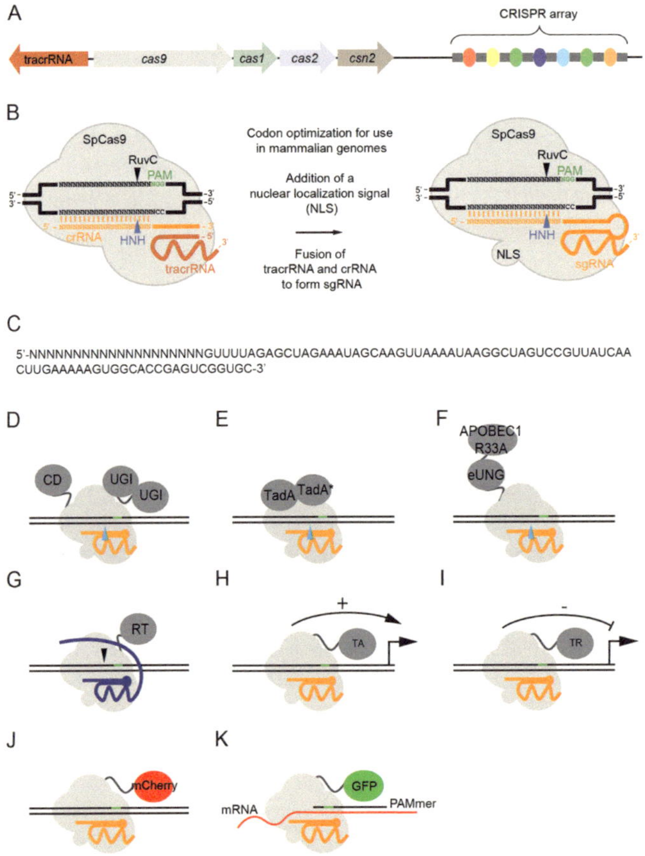

Figure 2.
CRISPR-Cas9. A) Schematic representation of the CRISPR-Cas9 operon from Streptococcus pyogenes. The operon contains 4 genes, three of which are involved in the adaptation stage of the immune response (Cas1, Cas2 and Csn2) and the maturation and interference stage of the response (Cas9). The locus also contains a tracrRNA and the CRISPR array which comprises repeat elements (gray rectangles) regularly interspaced with spacer elements (colored circles). B) Schematic representation of the CRISPR-SpCas9 system for genome editing and modifications made for its use in mammalian cells. The target genomic DNA is shown in black, with the PAM sequence (5′-NGG-3′) highlighted in green. The crRNA is shown in orange and the tracrRNA is shown in red. Fusion of the crRNA and tracrRNA forms the sgRNA, which is also shown in orange. The RuvC-like and HNH-like activity sites are shown by a black or a blue triangle, respectively. C) Sequence of the CRISPR-spCas9 sgRNA. The SpCas9 sgRNA contains sequences matching the protospacer element followed (N) by a hairpin loop linking the crRNA to the tracrRNA. D-K) schematic representation of various CRISPR_SpCas9 modalities including base editors (D-F), prime editors (G), target specific transcriptional activators or transcriptional repressors (H, I), as well as DNA and RNA tracking devices (J, K). D-F) base editors are formed from the fusion between catalytically impaired SpCas9 and base modifying enzymes such as CD which allows for the conversion of cytidine into a thymine (D), TadA-TadA heterodimer which allows for the conversion of adenine to guanine (E) and eUNG and APOBEC1 fusion which allows for the transversion of cytidine to guanine (F). G) Prime editors are produced by the combination of catalytically impaired SpCas9 and a reverse transcriptase (RT). Prime editors use a modified sgRNA called pegRNA that not only contains sequences providing target specifificty and Cas9 scaffolding, but also sequences complementary to the target sites and the substitutions to be engineered. Following the cleavage of the protospacer element, a 3′ DNA flap is exposed, which allows the binding of the pegRNA complementary segment to bind target site and serves as template for th RT. the RT will extend the 3′ flap, introducing the designed mutations. H, I) transcriptional modulators usually make use of transcriptional activators such as the viral transcription factor VP64. Transcriptional repressors typically make use of transcriptional repressors or epigenome modifiers such as Kruppel associated box (KRAB) MECP2 or DNMT3a. K) DNA tracking devices are formed from the fusion of a catalytically inactive SpCas9 fused to a fuuorescent marker. Tiling of multiple of these SpCAs9 fusions allows for the visualization of genomic loci in real time imaging. L) RNA tracking devices are formed form the coupling of a catalytically inactive SpCas9 to a fluorescent protein marker. Binding of SpCas9 with a target RNA transcript is promoted by the coadministration of a short oligonucleotide called PAMer that acts as exogenous PAM sequence.*

System	Bacterial strain	PAM or PFS
Cas9	Streptococcus pyogenes	NGG
	S. pyogenes (VQR)	NGAG
	S. pyogenes (VRER)	NGCG
	Streptococcus mutans	NGG
	Staphylococcus aureus	NNGGGT NNGAAT NNGAGT
	Streptococcus thermophilus (CRISPR3)	NGGNG
	S. thermophilus (CRISPR1)	NNAAAAW
	Campylobacter jejuni	NNNNACA
	Neisseria meningitidis	NNNNGATT
	Pasteurella multocida	GNNNCNNA
	Francisella novicida	NG
	Treponema denticola	NAAAAN
Cpf1	Franciselle novicida	TTN
	Acidaminococcus sp. BV3L6	TTTN
	Moraxella bovoculi 237	(T/C)(T/C)N
Cascade-Cas3	*Thermobifida fusca*	AAG
	Escherichia coli	ARG
	Pseudomonas aeruginosa	AAG
CRISPR-Cas12k	Scytonema hofmannii (ShCAST)	NGTN
	Anabaena cylindrica (AcCAST)	NGTN
Cascade-Tn6677	Vibrio Clolerae (Tn6677)	CC
CRISPR-Cas13	Leptotrichia shahii	A,U,C (not G)
CRISPR- Cas7–11	Scalindua brodae	N/A

PAM, protospacer adjacent motif; PFS, protospacer flanking site.

Table 1.
PAM or PFS requirements for various CRISPR-Cas systems.

the open reading frame of SpCas9 has been optimized for codon usage and nuclear localization signals (NLSs) have also been added.

CRISPR-SpCas9 systems have also been developed to introduce a limited number of nucleotide substitutions without introducing DNA DSBs. These systems, called Base Editors and Prime Editors, make use of catalytically impaired SpCas9 fused to various base modifying enzymes like a cytidine deaminase (CD), an uracil DNA N-glycosylase (eUNG), or a modified dimeric tRNA adenine deaminase (TadA*), which catalyze base transversion or base conversion within a precise window upstream of the PAM sequence (**Figure 2D-K**). The selection of a base editor depends on several criteria including the desired edit, the availability of PAM sequences within the target sequence, the position of the target nucleotide relative to the PAM sequence, the possibility of engineering undesired bystander mutations, and the need to minimize off-target editing [18]. The most recent versions of these systems are BE4max, a cytidine base editor which catalyzes the conversion of a cytidine into a thymine (C- > T) (**Figure 2D**); BE7.10, an adenine base editor which catalyzes the conversion of adenine to guanine (A- > G) (**Figure 2E**); CGBE1, a base editor that catalyzes cytidine to guanine (C- > G) base transversion (**Figure 2F**) [19–21].

Prime editing, on the other hand, makes use of a catalytically impaired SpCas9 fused to a reverse transcriptase (RT) and a prime editing guide RNA (pegRNA) (**Figure 2G**) [22]. The pegRNA not only provides target specificity and scaffolding, but also contains sequences that are complementary to the target site and substitutions encoding the desired edits. Following excision of the target strand, a 3′ flap is exposed and the pegRNA complexes with the exposed 3′ flap and serves as primer site for the RT, which extends the 3′ flap and incorporates the desired nucleotide substitutions. Stabilization of the locus is performed by the endogenous endonuclease FEN1 which removes the 5′ flap and allows the hybridization of the edited 3′ flap, resulting in the incorporation of edited bases and conversion of the unmodified allele via the DNA mismatch repair (MMR) pathway.

In addition to providing the framework for various genome editing technologies, CRISPR-SpCas9 has been morphed into DNA and RNA imaging devices [23, 24],

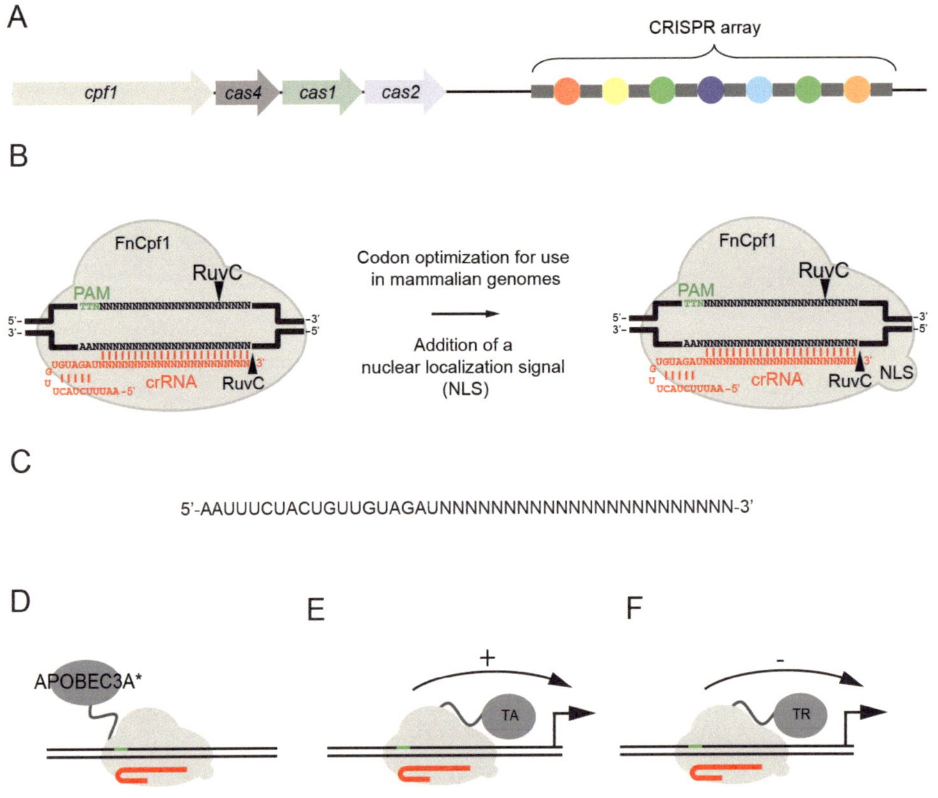

Figure 3.
CRISPR-Cpf1. A) Schematic representation of the CRISPR-Cpf1 operon from Francisella novicida. B) Schematic representation of the CRISPR-FnCpf1 system and the modifications for its usage in mammalian cells. The target genomic DNA is shown in black, with the PAM sequence (5′-TTN-3′) highlighted in green. The crRNA is shown in red, and the two RuvC-like activity sites are shown by black triangles. C) Sequence of the CRISPR-FnCpf1 crRNA. FnCpf1 crRNA. D) Cpf1 base editor. Like Cas9 base editors, Cpf1 base editors are from ed. from the fusion of Cpf1 with. E) Cpf1 transcriptional activator. F) Cpf1 transcriptional repressor.

epigenetic modifiers [25] as well as transcriptional modulators [26–28] via the fusion between catalytically inactive but sgRNA competent SpCas9 and transcriptional activators, transcriptional repressors, epigenetic modifiers, fluorescent proteins, and others (**Figure 2H-K**).

2.2 CRISPR-Cpf1

Another popular system for genome editing in mammalian cells has been derived from the Class 2 type V CRISPR-Cpf1 system from *F. novicida* (CRISPR-FnCpf1) (**Figure 3A**) [15]. In this system, a large endonuclease called FnCpf1 (or Cas12a) pairs with a 42 nucleotide-long crRNA that contains sequences providing both target specificity and FnCpf1 binding activity (**Figure 3B** and **C**). Unlike SpCas9, FnCpf1 contains two RuvC-like activities and introduces scattered DNA DSBs outside of its recognition sequence. More specifically, DNA breaks occur at positions 18 of the non-target strand and 23 of the target strand, leaving a 5′ overhang (**Figure 3B**). For FnCpf1 to be active, the endonuclease must also recognize a short PAM sequence (5′-TTN-3′) located at the 5′ end of the target sequence (or protospacer element). Similar to Cas9 systems, a variety of Cpf1 systems from diverse bacterial species with distinct PAM requirements have been identified, further expanding targeting possibilities. These include CRSIPR-Cpf1 systems from *Acidaminococcus sp. BV3L6*, *Lachnospiraceae bacterium MA2020*, and *Moraxella bovoculi 237*, among others [15]. Like CRISPR-SpCas9, CRISPR-FnCpf1 has also been converted into RNA-guided epigenetic modifying devices, transcriptional regulators and base editors (**Figure 3D-F**) [29–33].

3. Emerging CRISPR-Cas systems for genome and transcriptome modification

Concomitant with the development of CRISPR-Cas9 and CRISPR-Cpf1 for genome editing, other CRISPR-Cas operons have been adapted for genome and transcriptome modifications. These emerging technologies provide a means to engineer large genomic deletions, large DNA insertions into safe harbor loci, which remains somewhat challenging using conventional CRISPR-Cas systems, or for the modulation of gene expression. These systems include Cascade-Cas3 from *T. fusca*, an RNA-guided DNA helicase-nuclease capable of removing large DNA segments; CAST from *Scytonema hofmanni*, *Anabaena cylindrica* and *Vibrio cholerae* (Tn6677) which allow for RNA-guided insertion of large DNA cargos to any specified location within a genome, and RNA-guided RNA modifying systems from *L. shahii* or *Candidatus Scalindua brodae* for transcriptome silencing or editing.

3.1 Cascade-Cas3

The Type I CRISPR-Cas system from *T. fusca*, which has been used for genome editing of human cell lines, comprises a multi-subunit protein complex called Cascade that pairs with a short 61-nucleotide long crRNA to recruit a highly processive DNA helicase-nuclease named Cas3 (**Figure 4A-C**) [5]. The Cascade complex from *T. fusca* comprises six Cas7 subunits, two Cas11 subunits, as well as Cas5, Cas8 and Cas6 subunits (**Figure 4A**). Once bound to its target DNA, Cascade undergoes massive conformational changes stabilizing the newly formed R-loop (**Figure 4B**). This allows for the recruitment of Cas3 which then nicks the non-paired DNA strand

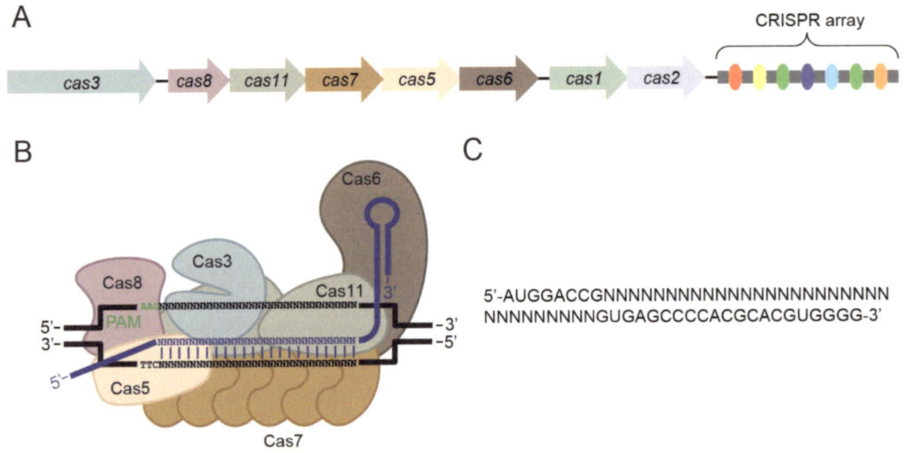

Figure 4.
Cascade-Cas3. A) Schematic representation of the Cascade-Cas3 operon from Thermobifida fusca. B) Schematic representation of the T. fusca Cascade-Cas3 system. The target genomic DNA is shown in black, with the PAM sequence (5′-AAG-3′) highlighted in green. The crRNA is shown in dark blue. The Cas3 subunit is shown in light blue, Cas8 shown in purple, Cas5 shown in tan, Cas6 shown in dark brown, two Cas11 subunits shown in green, and six Cas7 subunits shown in light brown. C) Sequence of the T. fusca Cascade-Cas3 crRNA.

and unidirectionally shreds, in a 3′ to 5′ orientation, the target DNA upstream of the PAM sequence (5′-AAG-3′). In this system, the PAM lies 5′ of the protospacer element. Cascade-Cas3 from *T. fusca* can generate long range deletions, from a few hundred to several thousand nucleotides. For editing in mammalian cells, sequences encoding NLSs were added to Cas3 and Cas7 subunits.

Several other Cascase-Cas3 systems have been developed for genome editing in bacteria and human cell lines. These include Type I-E Cascade-Cas3 from *E. coli*, and type I-C from *Pseudomonas aeruginosa* [4–6]. These systems have different PAM requirements, use slightly different Cas components, as well as crRNAs of different lengths and structures.

3.2 CRISPR-associated transposases

Insertion of large DNA elements using CRISPR-SpCas9 or CRISPR-FnCpf1 technologies has remained a major challenge. One emerging technology that may resolve this issue is CAST, which functions by recruiting Tn7-like transposase components to a specific location within the genome of a cell via guide RNA-target complementarity recognized by a naturally occurring inactive Cas12k variant. The Class 2 Type V-K Tn7-like CRISPR system from *Scytonema hofmanni*, the best described CAST system, comprises 6 components: Cas12k, a CRISPR-associated protein lacking endonuclease activity, a 216 nucleotide-long tracrRNA, a 34 nucleotide-long crRNA, and Tn7-like transposase subunits encoded by genes tnsB, tnsC, and tniQ (**Figure 5A** and **B**) [7, 16]. Similar to CRISPR-Cas9 systems adapted for genome editing, the type V-K tracrRNA and crRNA can be fused together to form a sgRNA. For Cas12k to recognize its target DNA, a 5′-NGTN-3′ PAM sequence must be present at the 5′ end of the protospacer element. Cargo insertion occurs unidirectionally in a 5′ Left End (LE) to 3′ Right End (RE) orientation (**Figure 5B, C**). A 5 bp integration site is found both 5′ and 3′ of the integration site. Cargos up to 10 kilobases can be introduced using this system and their integration into the host genome occurs 60–66 nucleotides downstream of the

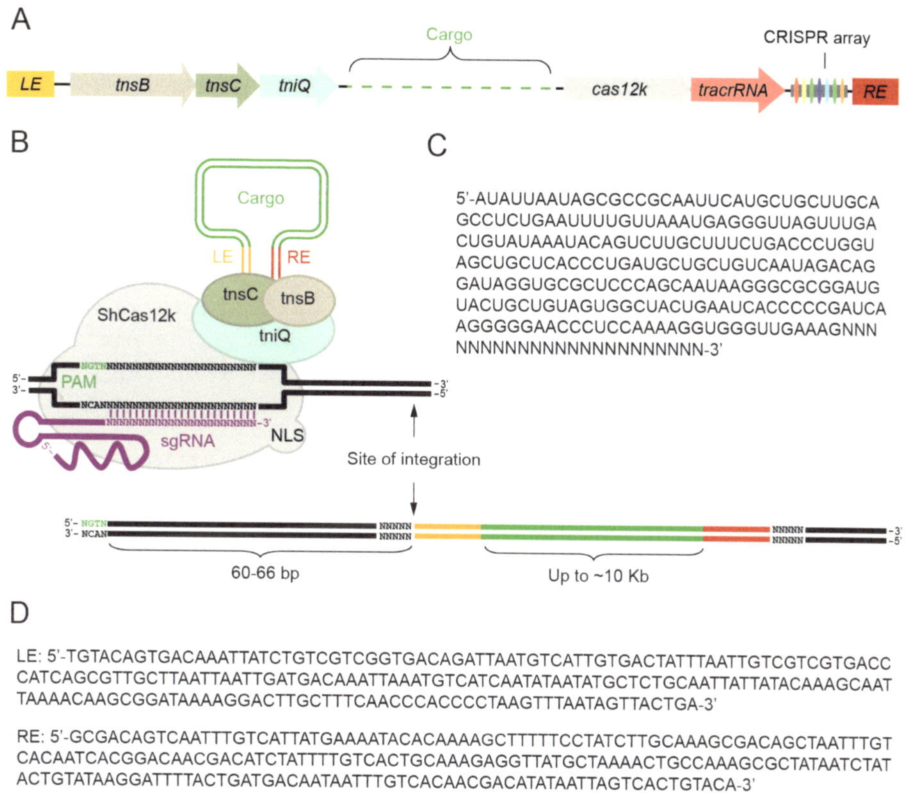

Figure 5.
CRISPR-Cas12k and associated transposase. A) Schematic representation of the CAST-Cas12k operon from Scytonema hofmanni. B) Schematic representation of the S. hofmanni CAST-Cas12 system. The target genomic DNA is shown in black, with the PAM sequence (5′-NGTN-3′) highlighted in green. The sgRNA is shown in purple. The tnsB, tnsC, and tniQ subunits are shown in gray, green, and light blue, respectively. The cargo DNA to be inserted, shown in green, is flanked by LE and RE sequences, shown in yellow and red, respectively. Black arrows indicate the site of integration, which occurs 60–66 bp downstream of the PAM. A 5 bp integration site is found both 5′ and 3′ of the integration site. C) Sequence of the S. hofmanni CAST-Cas12k sgRNA.Nucleotide sequence corresponds to the sgRNA for ShCas12k. Ns represent the CRISPR spacer. D) Sequences of the RE and LE from the S. hofmanni CAST-Cas12 system.

PAM sequence (**Figure 5C**). A variety of CAST systems have been identified in several bacterial species. These include CAST systems from *A. cylindrica* (AcCAST) which comprises similar components, possess similar PAM requirements (5′-NGTN-3′) and promotes insertion of transposons 49–56 nucleotides downstream of the PAM sequence [7, 16], as well as the Type I-C Cascade Tn7-like transposase system from *V. cholerae* (Tn6677), which promotes the bidirectional insertion of transposed elements 47–51 nucleotides downstream of PAM sequence [34].

3.3 RNA interference systems

Whereas the vast majority of CRISPR-Cas systems have evolved to protect against invading DNA species, Type VI CRISPR-Cas13 and the newly identified Type III-E CRISPR-Cas7–11 effectors are RNA-guided RNA interfering systems. These systems have been used to silence gene expression at the transcriptional level and have been modified to edit RNA transcripts. Several Type VI and Type III-E systems have been described. For simplicity, we present two of these systems: Type VI CRISPR-Cas13 from *L. shahii* (Cas13a) [9] and Type III-E CRISPR-Cas7–11 from Candidatus *Scalindua brodae* (Sb-gRAMP) [17].

Type VI CRISPR-Cas13a from *L. shahii* (also referred to as CRISPR-LshC2c2) is a single protein effector system that comprises a large ribonuclease containing two Higher Eukaryotes and Prokaryotes Nucleotide (HEPN) binding domains called C2c2 that pairs with a short 54 nucleotide-long crRNA to promote the cleavage of ssRNA transcripts at uracil residues [9] (**Figure 6A** and **B**). Of the 54 nucleotides, 28 residues provide target specificity (antisense to the protospacer) and the other 26 residues pair with C2c2 (**Figure 6C**). Cleavage of the ssRNA is sensitive to the nucleotide composition at the 3′end of the protospacer (also referred to as protospacer-flanking site or PFS). Spacers with a G immediately flanking the 3′ end of the protospacer were cleaved less efficiently than those containing any other nucleotides.

Not only have Type VI CRISPR-Cas13 systems been developed for the degradation of RNA species, but the fusion of a nuclease-dead Cas13b from *Prevotella sp. P5–125* with the adenosine deaminase acting on RNA type 2 (ADAR2) allows for the editing of RNA transcripts without interfering with the genomic sequence [8]. This may have applications for the treatment and understanding of genetic disorders.

The Type III-E effector system Cas7–11 from *Candidatus Scalindua brodae*, also referred to as giant Repeat-Associated Mysterious Protein (Sb-gRAMP), is a single-protein effector system that comprises a large modular protein containing four Cas7-like and a single Cas11-like domains with intrinsic endoribonuclease activity that pairs with a 47 nucleotide-long crRNA to cleave ssRNA at position 3 and 9 of the spacer [17] (**Figure 7A** and **B**). The 5′ most 27 nucleotides of the crRNA encode the direct repeat segment of the crRNA whereas the subsequent 20 nucleotides provide target specificity (**Figure 7C**). Several Class 1 type III-E systems have been identified, each requiring specific crRNA and endonuclease activity, most of which also introduce ssRNA breaks 6 nucleotides apart, at position 3 and 9 of the spacer [10]. Unlike DNA targeting systems, RNA targeting systems do not require PAM sequences, but their activity may be influenced by a PFS.

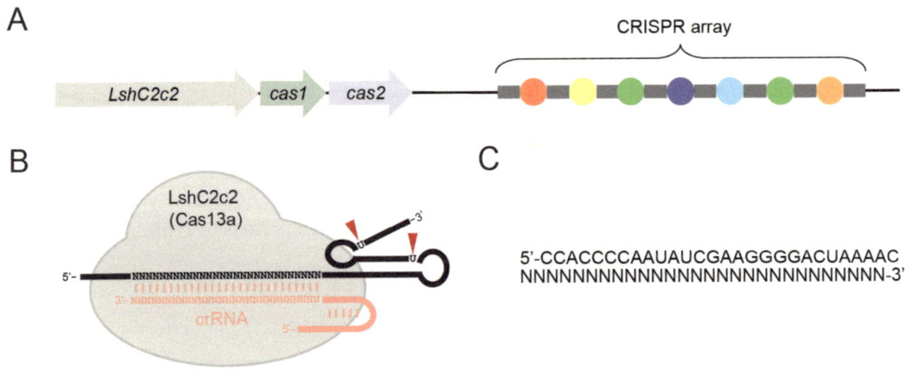

Figure 6.
CRISPR-Cas13. A) Schematic representation of the CRISPR-Cas13a (LshC2c2) operon from Leptotrichia shahii. B) Schematic representation of the CRISPR-Cas13a (LshC2c2) system. The target RNA is shown in black with cleavage sites indicated by black arrows. The crRNA is shown in pink. Unlike DNA targeting systems, RNA targeting systems do not require PAM sequences, but their activity may be influence by a PFS. LshC2c2 cuts its target RNA at accessible uracil residues (red triangles). C) Sequence of the CRISPR-Cas13a (LshC2c2) crRNA.

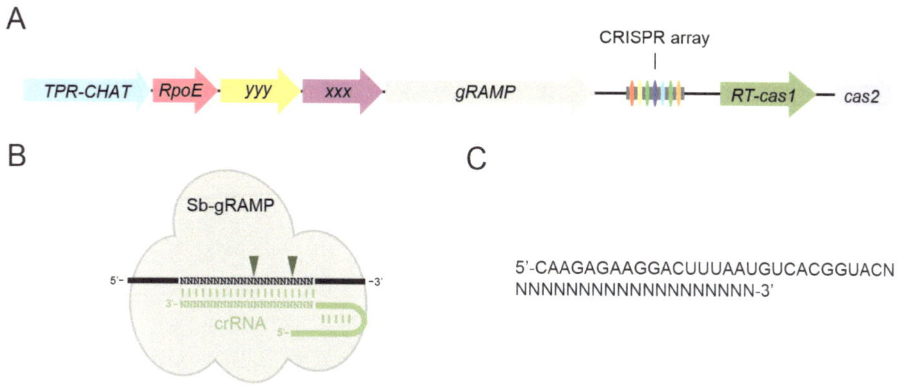

Figure 7.
CRISPR-Cas7–11. A) Schematic representation of the Cas7–11 operon from Candidatus Scalindua brodae. B) Schematic representation of the Cas7–11 Sb-gRAMP system. The target RNA is shown in black with cleavage sites indicated by black arrows 6 nucleotides apart, at position 3 and 9 of the spacer. The crRNA is shown in green. Unlike DNA targeting systems, RNA targeting systems do not require PAM sequences, but their activity may be influence by a PFS. C) Sequence of the Cas7–11 Sb-gRAMP crRNA.

4. Discussion

Whereas the vast majority of genetic manipulations can be performed using conventional CRISPR-SpCas9 technology, there are some inherent limitations that may be alleviated by emerging technologies. These include the possibility of introducing DNA DSBs at off-target sites; the possibility of inserting undesired mutations at on-target sites; the requirement for specific PAM sequences, which may somewhat limit the number of target possibilities; the scope of editing; and the delivery of these reagents, particularly for manipulations in vivo or for therapeutic interventions. **Table 2** explores the similarities and differences between the systems described in this chapter as well as their advantages and limitations.

4.1 Off-target mutations

One major limitation associated with the use of CRIPSR-SpCas9 technology is the potential for inserting genetic changes at sites other than the intended ones, also referred to as off-target sites. Off-target cleavage may occur due to the lack of SpCas9 specificity, which stems from the tolerance of the endonuclease for RNA–DNA mismatches, RNA bulges, or DNA bulges [12, 13]. Although there are still no simple and definitive guidelines defining SpCas9 specificity, the number and the position of mismatches relative to the PAM sequence are important. Whereas a single mismatch within the first 13 nucleotides upstream of the PAM sequence can abrogate SpCas9 activity, up to seven mismatches at the 5′ end of the guide sequence can be tolerated [12, 13]. To avoid off-target modifications, various strategies have been established. These include the development of bioinformatic tools to identify highly specific target sequences; the modification of SpCas9 to improve specificity or the duration of its action within cells; and the development of delivery formats and methods to limit the duration of SpCas9 activity [14]. Emerging RNA-guided RNA modifying systems may also help resolves this issue. These systems can modulate gene expression by targeting RNA transcripts rather than modifying genes at the DNA level [9, 11].

System	Source	Mechanism	Advantages	Limitations
CRISPR-Cas9	*Streptococcus pyogenes*	Cas9 pairs with a sgRNA to introduce a DNA DSB 3 nt upstream of the PAM. DSBs are resolved by NHEJ or HR	A well-established, conventional system; can be used to engineer a plethora of genetic modifications, ranging from nucleotide substitutions to chromosomal translocations; a single effector system which may be easier to use as opposed to multiprotein effector systems	Off-target editing of genomic DNA may occur; uses a long sgRNA; uses a large endonuclease which may be difficult to package in viral delivery systems; introduces DNA DSBs which may have deleterious effects if not resolved properly, particularly for therapeutic applications
CRISPR-Cpf1	*Francisella novicida*	Cpf1 pairs with a crRNA to introduce scattered DNA breaks at position 18 of the non-target strand and 23 of the target strand, leaving a 5′ overhang. DSBs are resolved by NHEJ or HR	Can be used to engineer a plethora of genetic modifications, ranging from nucleotide substitutions to chromosomal translocations; uses a different PAM than SpCas9, extending the range of possible target sites; a single effector system which may be easier to use as opposed to multiprotein effector systems; uses a short crRNA that may be easier to synthesize and deliver	Off-target editing of genomic DNA may occur; uses a large endonuclease (although smaller than SpCas9) which may be difficult to package in viral delivery systems; introduces DNA DSBs which may have deleterious effects if not resolved properly, particularly for therapeutic applications
Cascade-Cas3	*Thermobifida fusca*	Cascade pairs with a crRNA to recruit Cas3, a highly processive DNA helicase-nuclease, which nicks the non-paired DNA strand and unidirectionally shreds the target DNA upstream of the PAM sequence in a 3′ to 5′ orientation	Can be used to generate large deletions; uses a short crRNA that may be easier to synthesize and deliver	Off-target editing of genomic DNA may occur; large multiprotein effector which can be difficult to package for viral delivery; limited in the scope of editing; no control over the length of the deletions
CRISPR-CAST	*Scytonema hofmanni*	Cas12k and Tn7-like transposase components pair with a sgRNA to insert large DNA cargo unidirectionally in a 5′ LE to 3′ RE orientation 60–66 nt downstream of the PAM. A 5 bp integration site is duplicated and found 5′ and 3′ of the cargo integration	Can be used to insert large DNA elements up to ~10Kb	Off-target editing of genomic DNA may occur; insertion results in the integration of flanking sequences and duplication of the integration site; a multiprotein effector which can be difficult to package for viral delivery; uses a long sgRNA; limited in the scope of editing; demonstrated efficacy only in bacterial systems

System	Source	Mechanism	Advantages	Limitations
CRISPR-Cas13	*Leptotrichia shahii*	C2c2, a HEPN containing protein, pairs with a crRNA to cleave ssRNA transcripts at exposed uracil residues	Targets RNA rather than DNA, can be used to silence gene expression at the translational level instead of permanently altering the DNA; a single effector system which may be easier to use as opposed to multiprotein effector systems; uses a short crRNA that may be easier to synthesize and deliver	Off-target editing of ssRNA transcripts may occur; cleavage of ssRNA transcripts occurs at exposed uracil residues, which are difficult to predict; a large effector protein which can be difficult to package for viral delivery; demonstrated efficacy only in bacterial systems
Cas7–11	*Candidatus Scalindua brodae*	A protein (gRAMP) with Cas7-like and Cas11-like domains with intrinsic endoribonuclease activity pairs with a crRNA to cleave ssRNA transcripts at positions 3 and 9 of the spacer	Targets RNA rather than DNA, can be used to silence gene expression at the translational level instead of permanently altering the DNA; cleavage of ssRNA transcripts occurs at positions 3 and 9 of the spacer, which are easily predictable compared to CRISPR-Cas13 systems; a single effector system which may be easier to use as opposed to multiprotein effector systems; uses a short crRNA that may be easier to synthesize and deliver	Off-target editing of ssRNA transcripts may occur; a large effector protein which can be difficult to package for viral delivery; demonstrated efficacy only in bacterial systems

Table 2.
Mechanisms, advantages, and limitations associated with prototypical CRISPR-Cas systems.

To identify highly selective guide sequences, various guide selection applications have been developed. These include Cas-Designer, quick guide-RNA designer for CRISPR/Cas derived RNA guided nucleases (http://www.rgenome.net/); CRISPR Design (http://crispr.mit.edu/);E-CRISP (http://www.e-crisp.org/E-CRISP/); ZiFit (http://zifi t.partners.org/ZiFiT/). During the implementation of CRISPR-SpCas9 technology for mouse genome editing, our laboratory also developed a stringent guide selection procedure which makes use of Cas-Designer and Cas-Offinder from CRISPR RGEN Tools (http://www.rgenome.net/). For more details about this guide selection procedure, we recommend reading.

To control the duration of SpCas9 activity within cells, genetically encoded inducible systems have been developed. One of these makes use of split SpCas9 fused to the Magnet Photoactivatable System. Fusion of the split SpCas9 is achieved by illuminating cells with blue light. The fused split SpCas9 is then able to bind a sgRNA and cleave its target site. A second system involves self-cleavable CRIPSR systems, where sequences encoding SpCas9 are targeted by a sgRNA in order to promote its degradation upon expression of the endonuclease. Duration of SpCas9 activity can also be controlled by the delivery of RNA transcripts encoding the various components of the CRISPR-SpCas9 system or via delivery of the ribonucleoprotein complex comprising both the sgRNA and the endonuclease.

In addition to robust guide selection procedures and inducible/self-inactivating systems, various additional strategies have been developed to improve target specificity. These include the use of paired SpCas9 nickases, in which the RuvC-like domains are inactivated, that introduce scattered DNA DSB, guided by a pair of sgRNAs recognizing juxtaposed sequences. The requirement for recognizing these sequences doubles the length of the target sequence and thus increases target specificity. Similarly, catalytically inactive but sgRNA competent pairs of SpCas9 fused to the non-specific endonuclease FokI was shown to reduce off-target activity. Directed evolution has also been used to engineer improved SpCas9 with increased target specificity. These enzymes have been shown to increase on-target over off-target activity by several folds.

Finally, modifications to the guide RNAs themselves were also used to reduce off-target cleavage. Previous studies have shown that, counterintuitively, shortened guide sequences increase specificity without affecting on-target activity.

4.2 Unintended mutations at the target site

Engineering specific mutations using conventional CRIPSR-SpCas9 technology relies on HR. While HR and NHEJ are both active in most (but not all) dividing cells, NHEJ is usually the sole repair pathway active in postmitotic cells. DNA repair via the NHEJ pathway, as previously mentioned, results in the insertion or deletion of genetic material, and does not allow for the introduction of desired mutations. To get around this, Base Editors and Prime Editors were developed. These systems, as described above, can introduce specific mutations by directly changing the nucleotide composition at the target site, bypassing the need to activate DSB repair pathways. Stabilization of the mutation is performed by the DNA MMR pathway which is present and active in all cells. While these systems allow for the insertion of precise mutations, the window in which they operate is narrow (a few nucleotides), the insertion of mutation(s) depends on the presence of a PAM sequence, stabilization of the mutation is not always complete, and undesired collateral nucleotide substitutions may occur [18–21, 31, 33]. Moreover, these systems are bulkier than SpCas9 and may not be easily packaged within viral delivery systems. Nevertheless, these systems provide an alternative to conventional CRISPR-SpCas9 systems for engineering mutations in cells that are not amenable to HR or for therapeutic intervention where introducing DNA DSBs may have deleterious effects.

4.3 Editing scope

While conventional CRISPR-Cas9 systems can be used to engineer virtually any kind of mutations in vivo, insertions or deletions of large DNA elements remain somewhat challenging. Cascade-Cas3 and CRISPR-CAST systems may provide alternatives to using conventional CRISPR-Cas9 systems. Cascade-Cas3 systems have been used to engineer large deletions in cultured cells and can potentially be applied to animal models [4–6]. Deletions range from several hundred to several thousand base pairs. The major drawback of using this technology is the apparent uncontrollable processivity of the helicase-nuclease Cas3. Consequently, deletions of various sizes must be characterized using a large number of primer pairs flanking the potential deletions. Other limitations include the difficulty of packaging multiprotein systems for viral delivery in vivo. CRISPR-CAST systems, on the other hand, allow for targeted integration of large genetic material. In *E. coli*, up to 10 kilobases of DNA have been successfully inserted. Unlike conventional CRISPR-Cas9 systems, however, which allow for scarless integration of genetic material, insertion of DNA elements using CRISPR-CAST systems results in the integration of flanking sequences (LE and RE) as well as duplication of the integration site. Consequently, this technology cannot be used for precise DNA insertion, but can be used to facilitate insertion of transgenes at safe harbor sites within genomes. Like other multiprotein effector systems, CRISPR-CAST systems may also be difficult to package for viral delivery in vivo. Although CRISPR-CAST systems have only been used in bacteria, the implementation of these systems for their use in mammalian genome engineering is likely and provides an alternative to conventional CRISPR-Cas9 systems.

4.4 Delivery

The vast majority of CRISPR-Cas systems for genome editing make use of either multiprotein effectors or large single effectors. Although delivering these systems, together with their cognate guide RNAs, in cultured cells or zygotes for the generation of animal models does not represent a major hurdle and is routinely performed, delivering these systems in vivo, for therapeutic interventions, does represent a major challenge. More compact CRISPR-Cas9 and Cas3 systems have been identified and these may represent viable alternatives to other larger and more complex CRISPR-Cas system for therapeutic purposes [6].

4.5 PAM requirements

The requirement for CRISPR-Cas systems to recognize short genomic sequences has long been viewed as major disincentive for the use of these technologies for genome engineering. However, most PAM sequences are quite short and are likely present at a high frequency within mammalian genomes. Moreover, the development of several CRISPR-Cas9 and Cpf1 systems with distinct PAM requirements and the generation of engineered SpCas9 endonucleases with altered PAM specificities have expanded the targeting capabilities of CRISPR-Cas systems.

5. Conclusion

Since the discovery of targetable nucleases, more notably CRISPR-Cas systems, the field of genetic and genome engineering has expanded exponentially. In less than a decade, these systems have not only revolutionized how research is performed but have also allowed for a plethora of scientific discoveries and paved the way for novel human therapeutics. Emerging technologies such as Cascade-Cas3, CAST, CRISPR-Cas7–11, and CRISPR-Cas13 provide alternatives to current technologies and may fill a critical technological gap to improve the specificity and scope of genome editing. Moreover, the implementation of these tools as therapeutic agents offers the potential to treat or even cure human genetic diseases.

Acknowledgements

The author would like to thank the Indiana University School of Medicine for the financial support of the Indiana University Genome Editing Center.

Author details

Annelise Cassidy and Stephane Pelletier*
Indiana University School of Medicine, Indianapolis, IN, United States of America

*Address all correspondence to: spellet@iu.edu

References

[1] Christian M, Cermak T, Doyle EL, Schmidt C, Zhang F, Hummel A, et al. Targeting DNA double-strand breaks with TAL effector nucleases. Genetics. 2010;**186**(2):757-761. DOI: 10.1534/genetics.110.120717

[2] Bibikova M, Beumer K, Trautman JK, Carroll D. Enhancing gene targeting with designed zinc finger nucleases. Science. 2003;**300**(5620):764. DOI: 10.1126/science.1079512

[3] Makarova KS, Wolf YI, Alkhnbashi OS, Costa F, Shah SA, Saunders SJ, et al. An updated evolutionary classification of CRISPR-Cas systems. Nature Reviews. Microbiology. 2015;**13**(11):722-736. DOI: 10.1038/nrmicro3569

[4] Morisaka H, Yoshimi K, Okuzaki Y, Gee P, Kunihiro Y, Sonpho E, et al. CRISPR-Cas3 induces broad and unidirectional genome editing in human cells. Nature Communications. 2019;**10**(1):5302. DOI: 10.1038/s41467-019-13226-x

[5] Dolan AE, Hou Z, Xiao Y, Gramelspacher MJ, Heo J, Howden SE, et al. Introducing a Spectrum of long-range genomic deletions in human embryonic stem cells using type I CRISPR-Cas. Molecular Cell. 2019;**74**(5):936-50 e5. DOI: 10.1016/j.molcel.2019.03.014

[6] Csorgo B, Leon LM, Chau-Ly IJ, Vasquez-Rifo A, Berry JD, Mahendra C, et al. A compact Cascade-Cas3 system for targeted genome engineering. Nature Methods. 2020;**17**(12):1183-1190. DOI: 10.1038/s41592-020-00980-w

[7] Strecker J, Ladha A, Gardner Z, Schmid-Burgk JL, Makarova KS, Koonin EV, et al. RNA-guided DNA insertion with CRISPR-associated transposases. Science. 2019;**365**(6448):48-53. DOI: 10.1126/science.aax9181

[8] Cox DBT, Gootenberg JS, Abudayyeh OO, Franklin B, Kellner MJ, Joung J, et al. RNA editing with CRISPR-Cas13. Science. 2017;**358**(6366):1019-1027. DOI: 10.1126/science.aaq0180

[9] Abudayyeh OO, Gootenberg JS, Konermann S, Joung J, Slaymaker IM, Cox DB, et al. C2c2 is a single-component programmable RNA-guided RNA-targeting CRISPR effector. Science. 2016;**353**(6299):aaf5573. DOI: 10.1126/science.aaf5573

[10] Kato K, Zhou W, Okazaki S, Isayama Y, Nishizawa T, Gootenberg JS, et al. Structure and engineering of the type III-E CRISPR-Cas7-11 effector complex. Cell. 2022;**185**(13):2324-37 e16. DOI: 10.1016/j.cell.2022.05.003

[11] Ozcan A, Krajeski R, Ioannidi E, Lee B, Gardner A, Makarova KS, et al. Programmable RNA targeting with the single-protein CRISPR effector Cas7-11. Nature. 2021;**597**(7878):720-725. DOI: 10.1038/s41586-021-03886-5

[12] Hsu PD, Scott DA, Weinstein JA, Ran FA, Konermann S, Agarwala V, et al. DNA targeting specificity of RNA-guided Cas9 nucleases. Nature Biotechnology. 2013;**31**(9):827-832. DOI: 10.1038/nbt.2647

[13] Jinek M, Chylinski K, Fonfara I, Hauer M, Doudna JA, Charpentier E. A programmable dual-RNA-guided DNA endonuclease in adaptive bacterial immunity. Science. 2012;**337**(6096):816-821. DOI: 10.1126/science.1225829

[14] Mali P, Yang L, Esvelt KM, Aach J, Guell M, DiCarlo JE, et al. RNA-guided human genome engineering via Cas9. Science. 2013;**339**(6121):823-826. DOI: 10.1126/science.1232033

[15] Zetsche B, Gootenberg JS, Abudayyeh OO, Slaymaker IM, Makarova KS, Essletzbichler P, et al. Cpf1 is a single RNA-guided endonuclease of a class 2 CRISPR-Cas system. Cell. 2015;**163**(3):759-771. DOI: 10.1016/j.cell.2015.09.038

[16] Strecker J, Jones S, Koopal B, Schmid-Burgk J, Zetsche B, Gao L, et al. Engineering of CRISPR-Cas12b for human genome editing. Nature Communications. 2019;**10**(1):212. DOI: 10.1126/science.aax9181

[17] van Beljouw SPB, Haagsma AC, Rodriguez-Molina A, van den Berg DF, Vink JNA, Brouns SJJ. The gRAMP CRISPR-Cas effector is an RNA endonuclease complexed with a caspase-like peptidase. Science. 2021;**373**(6561):1349-1353. DOI: 10.1126/science.abk2718

[18] Anzalone AV, Koblan LW, Liu DR. Genome editing with CRISPR-Cas nucleases, base editors, transposases and prime editors. Nature Biotechnology. 2020;**38**(7):824-844. DOI: 10.1038/s41587-020-0561-9

[19] Komor AC, Kim YB, Packer MS, Zuris JA, Liu DR. Programmable editing of a target base in genomic DNA without double-stranded DNA cleavage. Nature. 2016;**533**(7603):420-424. DOI: 10.1038/nature17946

[20] Gaudelli NM, Komor AC, Rees HA, Packer MS, Badran AH, Bryson DI, et al. Programmable base editing of a*T to G*C in genomic DNA without DNA cleavage. Nature. 2017;**551**(7681):464-471. DOI: 10.1038/nature24644

[21] Kurt IC, Zhou R, Iyer S, Garcia SP, Miller BR, Langner LM, et al. CRISPR C-to-G base editors for inducing targeted DNA transversions in human cells. Nature Biotechnology. 2021;**39**(1):41-46. DOI: 10.1038/s41587-020-0609-x

[22] Anzalone AV, Randolph PB, Davis JR, Sousa AA, Koblan LW, Levy JM, et al. Search-and-replace genome editing without double-strand breaks or donor DNA. Nature. 2019;**576**(7785):149-157. DOI: 10.1038/s41586-019-1711-4

[23] Chen B, Gilbert LA, Cimini BA, Schnitzbauer J, Zhang W, Li GW, et al. Dynamic imaging of genomic loci in living human cells by an optimized CRISPR/Cas system. Cell. 2013;**155**(7):1479-1491. DOI: 10.1016/j.cell.2013.12.001

[24] Nelles DA, Fang MY, O'Connell MR, Xu JL, Markmiller SJ, Doudna JA, et al. Programmable RNA tracking in live cells with CRISPR/Cas9. Cell. 2016;**165**(2):488-496. DOI: 10.1016/j.cell.2016.02.054

[25] Vojta A, Dobrinic P, Tadic V, Bockor L, Korac P, Julg B, et al. Repurposing the CRISPR-Cas9 system for targeted DNA methylation. Nucleic Acids Research. 2016;**44**(12):5615-5628. DOI: 10.1093/nar/gkw159

[26] Qi LS, Larson MH, Gilbert LA, Doudna JA, Weissman JS, Arkin AP, et al. Repurposing CRISPR as an RNA-guided platform for sequence-specific control of gene expression. Cell. 2013;**152**(5):1173-1183. DOI: 10.1016/j.cell.2013.02.022

[27] Gilbert LA, Horlbeck MA, Adamson B, Villalta JE, Chen Y, Whitehead EH, et al. Genome-scale CRISPR-mediated control of gene repression and activation. Cell. 2014;**159**(3):647-661. DOI: 10.1016/j.cell.2014.09.029

[28] Cheng AW, Wang H, Yang H, Shi L, Katz Y, Theunissen TW, et al. Multiplexed activation of endogenous genes by CRISPR-on, an RNA-guided transcriptional activator system. Cell Research. 2013;**23**(10):1163-1171. DOI: 10.1038/cr.2013.122

[29] Zhang X, Wang J, Cheng Q, Zheng X, Zhao G, Wang J. Multiplex gene regulation by CRISPR-ddCpf1. Cell Discovery. 2017;**3**:17018. DOI: 10.1038/celldisc.2017.18

[30] Liu Y, Han J, Chen Z, Wu H, Dong H, Nie G. Engineering cell signaling using tunable CRISPR-Cpf1-based transcription factors. Nature Communications. 2017;**8**(1):2095. DOI: 10.1038/s41467-017-02265-x

[31] Li X, Wang Y, Liu Y, Yang B, Wang X, Wei J, et al. Base editing with a Cpf1-cytidine deaminase fusion. Nature Biotechnology. 2018;**36**(4):324-327. DOI: 10.1038/nbt.4102

[32] Tak YE, Kleinstiver BP, Nunez JK, Hsu JY, Horng JE, Gong J, et al. Inducible and multiplex gene regulation using CRISPR-Cpf1-based transcription factors. Nature Methods. 2017;**14**(12): 1163-1166. DOI: 10.1038/nmeth.4483

[33] Wang X, Ding C, Yu W, Wang Y, He S, Yang B, et al. Cas12a base editors induce efficient and specific editing with low DNA damage response. Cell Reports. 2020;**31**(9):107723. DOI: 10.1016/j.celrep.2020.107723

[34] Klompe SE, Vo PLH, Halpin-Healy TS, Sternberg SH. Transposon-encoded CRISPR-Cas systems direct RNA guided DNA integration. Nature. 2019;**571**(7764):219-225. DOI: 10.1038/s41586-019-1323-z

Section 2

Technology

Chapter 3

Maximizing the Efficacy of CRISPR/Cas Homology-Directed Repair Gene Targeting

Terry S. Elton, Md. Ismail Hossain, Jessika Carvajal-Moreno, Xinyi Wang, Dalton J. Skaggs and Jack C. Yalowich

Abstract

Clustered regularly interspaced short palindromic repeats/CRISPR-associated system (CRISPR/Cas) is a powerful gene editing tool that can introduce double-strand breaks (DSBs) at precise target sites in genomic DNA. In mammalian cells, the CRISPR/Cas-generated DSBs can be repaired by either template-free error-prone end joining (e.g., non-homologous end joining/microhomology-mediated end joining [NHEJ]/[MMEJ]) or templated error-free homology-directed repair (HDR) pathways. CRISPR/Cas with NHEJ/MMEJ DNA repair results in various length insertions/deletion mutations (indels), which can cause frameshift mutations leading to a stop codon and subsequent gene-specific knockout (i.e., loss of function). In contrast, CRISPR/Cas with HDR DNA repair, utilizing an exogenous repair template harboring specific nucleotide (nt) changes, can be employed to intentionally edit out or introduce mutations or insertions at specific genomic sites (i.e., targeted gene knock-in). This review provides an overview of HDR-based gene-targeting strategies to facilitate the knock-in process, including improving gRNA cleavage efficiency, optimizing HDR efficacy, decreasing off-target effects, suppressing NHEJ/MMEJ activity, and thus expediting the screening of CRISPR/Cas-edited clonal cells.

Keywords: CRISPR/Cas, homology-directed repair, gene editing, Cas9, Cas12, non-homologous end joining, microhomology-mediated end joining, knock-in

1. Introduction

Clustered regularly interspaced short palindromic repeats/CRISPR-associated system (CRISPR/Cas) technology has revolutionized biological research and holds great therapeutic potential, since it is remarkably flexible and reliable [1–3]. CRISPR/Cas genome editing (i.e., genetic engineering) is a programmable technology to introduce double-strand breaks (DSBs) at specific target sites in the genome of a living organism [1–3]. There are two major mechanisms by which Cas enzyme-mediated DSBs are subsequently repaired [4–6]. The first is by template-free end joining (e.g., non-homologous end joining/microhomology-mediated end joining [NHEJ]/[MMEJ]), which introduces insertions/deletion mutations (indels) and can lead to targeted gene knock outs. The second mechanism is *via* the homology-directed repair (HDR) pathway, which produces a targeted gene knock-in or other specific mutations utilizing an exogenous donor template [4–6]. Given that DSBs generated in mammalian cells are predominantly repaired by NHEJ or MMEJ, the rate of precise editing through CRISPR/Cas/HDR with an exogenous repair template is significantly compromised/reduced [4–6]. This review summarizes multiple strategies to enhance the efficacy of CRISPR/Cas/HDR as well as decrease off-target effects.

2. CRISPR/Cas history

CRISPR history began in 1987 when Ishino et al. [7–9] first observed five repetitive palindromic sequences of 29 nucleotides separated by random 32 nucleotides toward the end of the *E. coli* genome. Although Ishino et al. [7] did not decipher the biological significance of the puzzling repeat sequences, this report led to the discovery of similar patterns in other bacterial and archaea genomes [10–12]. Mojica et al. [13] then established that the unusual repetitive DNA sequences were functionally related. These curious sequences were later designated "CRISPR" by Jansen et al. [14] given that these loci harbor: 1) palindromic repeats with little sequence variation; 2) non-repetitive spacer sequences between the repeats; and 3) a several hundred base pair (bp) common leader sequence on one side of the repeat cluster. The CRISPR locus is present in approximately 40% of the sequenced bacteria and 90% of the genomes of the different domains of archaea [15]. Finally, it was demonstrated that CRISPR-associated (Cas) genes (i.e., over 40), of which only a subset is found in any given prokaryote that harbor CRISPRs, are frequently located in close proximity to CRISPR loci [14, 16, 17]. The Cas genes were predicted to encode endo- and exonucleases, helicases, polymerases, and RNA-binding proteins [14, 15, 17].

Initially, CRISPR/Cas systems were expected to have a role in DNA repair or gene regulation due to their location near the DNA repair system in the bacterial genome [18]. However, in 2005, three seminal studies revealed that the CRISPR spacer sequences were homologous to bacteriophage, prophages, and conjugative plasmid sequences and suggested that they were the remnants of past invasions by extrachromosomal elements [19–21]. These investigators further speculated that there was a relationship between CRISPR and immunity against foreign DNAs by coding an anti-sense RNA [19–21]. In the following year, Makrova et al. [17] analyzed the link between the CRISPR and the Cas proteins and how this system is similar to the prokaryotic RNAi-mediated adaptive immune system, which led them to propose that the CRISPR/Cas system, with its "memory component," may function as inheritable adaptive immunity for bacteria.

Subsequently, Barrangou et al. [22] demonstrated that after a viral challenge, phage sequence was integrated into a CRISPR locus of *Streptococcus thermophilus* and provided immunity against the corresponding phage. When the protospacer sequence was deleted from the bacterial genome, they became sensitive to phage infection [22]. These investigators hypothesized that the nucleic acid based "immunity" system in prokaryotes was dictated by the CRISPR spacer sequence and that the Cas protein machinery mediated resistance against foreign DNAs [22].

In 2008, a pivotal study by Brouns et al. [23] established that the *E. coli* spacer sequences were transcribed into a precursor CRISPR RNA (pre-crRNA) that was matured to small crRNAs by a complex of Cas proteins. Additionally, it was demonstrated that mature crRNAs serves as a "guide" to a direct a protein to target viral nucleic acids, which results in an antiviral response in prokaryotes [23]. Subsequently, Mojica et al. [24] identified CRISPR-type-specific proto-spacer adjacent motifs (PAMs), which are important for discrimination between self and nonself sequences. Further, Garneau et al. [25] showed that CRISPR/Cas immunity resulted from the generation of DSBs at specific sites in bacteriophage and plasmid DNA. Finally, Sapranauskas et al. [26] demonstrated that the *S. thermophilus* CRISPR/Cas system could be transferred to *E. coli* and provide a Cas9-mediated immunity that required a PAM site. Their initial characterization of the Cas9 protein revealed that two domains are involved in the formation of DBSs [26]. The Cas9 McrA/HNH-like nuclease domain cleaves the DNA strand complementary to the guide RNA sequence (target strand), and the RuvC/RNaseH-like domain cleaves the noncomplementary strand (nontarget strand) (**Figure 1**) [26]. They also demonstrated that a 20-nucleotide crRNA, the trimmed version of the full-length crRNA, is sufficient for DNA target identification with efficient cleavage and that the target site of the Cas can be changed by changing the crRNA sequence (**Figure 1**) [26].

Next, Deltcheva et al. [27] discovered an additional small RNA designated the trans-activating CRISPR RNA (tracrRNA). This small RNA is transcribed from

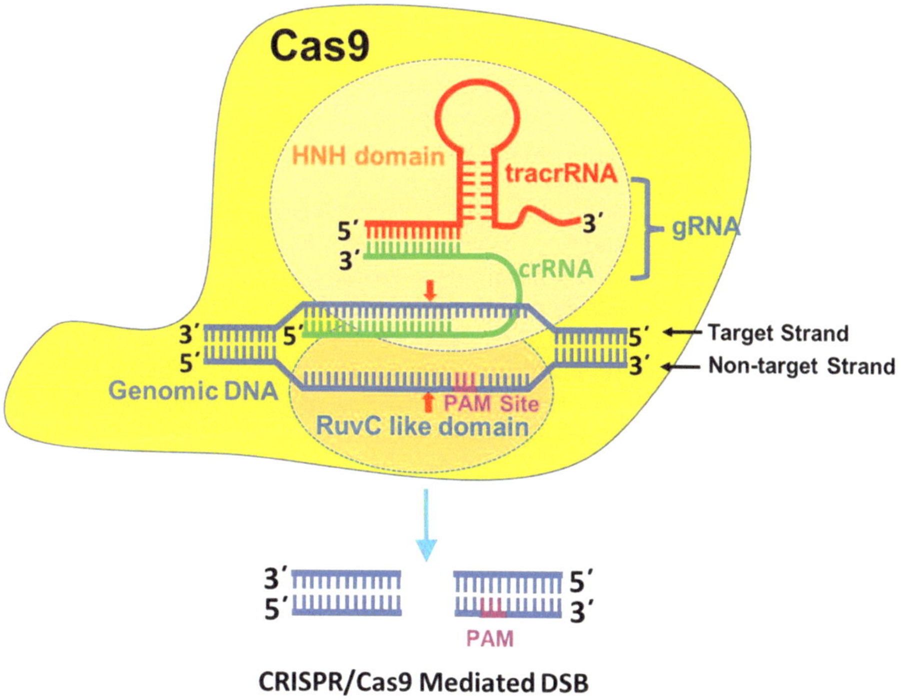

Figure 1.
Schematic representation of the CRISPR/Cas9-mediated DSB with a two-piece gRNA. The Cas9 gRNA is a two-piece RNA complex comprised of a crRNA required for DNA targeting (denoted in green) and the tracrRNA, which is necessary for nuclease activity (denoted in red) [27, 28]. The Cas9 protein (denoted in yellow) binds to the gRNA to form a RNP complex. The gRNA directs the Cas9 to a specific location in the genomic DNA (denoted in blue) through a user-defined 20 nt sequence at the 5′ end of the crRNA, which is complementary to the target DNA (denoted by green and blue hash marks). If there is a PAM site (NGG) (denoted in pink) adjacent to the 3′ end of the 20 nt sequence, then the Cas9 McrA/HNH-like nuclease domain (denoted in peach) cleaves the DNA strand complementary to the guide RNA sequence (target strand) and the RuvC/RNaseH-like domain (denoted in orange) cleaves the noncomplementary strand (nontarget strand) to introduce site-specific DSBs in the target DNA [26].

sequence upstream of the CRISPR-Cas locus of *Streptococcus pyogenes* [27]. These investigators demonstrated that, upon maturation of both the tracrRNA and the crRNA, they form a duplex that has both single- and double-stranded regions (**Figure 1**) [27]. Furthermore, Jinek et al. [28] verified that the two-RNA complex, dual RNA (i.e., the crRNA [required for DNA targeting] and the tracrRNA [necessary for nuclease activity]; now designated as a guide-RNA [gRNA]) directs the Cas9 to introduce site-specific DSBs in the target DNA (**Figure 1**). They also demonstrated that Cas9 target recognition required complementary seed sequences between the crRNA and target DNA as well as a PAM sequence containing a GG dinucleotide adjacent to the crRNA-binding region in the DNA target (**Figure 1**) [28]. Moreover, Jinek et al. [28] established that the *S. pyogenes* Cas9 endonuclease could be programmed to target and cleave any dsDNA sequence, which harbors a NGG (N denotes any nt) PAM site, with an engineered gRNA which contains a 20-nucleotide crRNA sequence that is complementary to the target DNA (**Figure 1**). CRISPR-Cas technology is now widely adopted in the scientific community due to its simplicity and precision for gene editing, which has opened the possibility of numerous applications in the field of genetic engineering.

3. DNA double-strand break repair

Pathological DNA DSBs can arise from normal endogenous metabolic cellular processes (e.g., DNA replication and transcription) or from cellular exposure to exogenous sources (e.g., reactive oxygen species, ionizing radiation, radiomimetic chemicals, and anticancer chemotherapeutic drugs) [29–32]. However, physiologically important DNA DSBs are also required for several developmental and physiological cellular activities including chromosomal disjunction, meiosis, V(D)J, and immunoglobulin heavy chain (IgH) class switch recombination [29, 33]. Notably, both pathological and physiological DNA DSBs require efficient repair processes since these lesions can result in insertions, deletions, chromosomal translocations, and genomic instability, which can lead to numerous hereditary human diseases, including cancer, developmental disorders, and premature aging [29–31, 33]. Mammalian cells employ multiple DNA repair pathways to protect the integrity of their genomes. However, the two predominate DNA DSB repair pathways that are template-free NHEJ/MMEJ and templated HDR [32, 34, 35]. It is important to note that NHEJ and HDR are two competing pathways [4–6]. In mammalian cells, template-free NHEJ is favored over templated HDR since NHEJ is a rapid high-capacity pathway, which is active throughout the cell cycle and directly represses HDR [4–6]. In contrast, HDR is largely restricted to the S and G2 phases [4–6].

At the most basic level, the CRISPR/Cas genome editing technology is utilized to introduce a DSB at a specific target site in the genome and then relies upon the cellular machinery to repair this lesion by either the NHEJ/MMEJ or HDR repair pathways to yield the desired repair outcomes [32, 34, 35]. If the experimental goal is to knockout the function of a given gene of interest, then the error-prone NHEJ or MMEJ pathways would be utilized to repair DNA DSBs created by the Cas endonuclease at a programmed target site to introduce indels, which can shift the open reading frame (ORF) and result in targeted gene loss of function [32, 34, 35]. In contrast, if the experimental goal is to edit out or to introduce mutations at specific genomic sites (i.e., targeted gene knock-in), then the HDR pathway would be utilized to repair the Cas endonuclease-created DNA DSBs with an exogenous repair template harboring specific nucleotide (nt) changes [32, 34, 35]. The cellular NHEJ/MMEJ and HDR repair pathways of endogenous and CRISPR/Cas-generated DSBs will be discussed in more detail below.

3.1 Template-free error-prone end joining NHEJ/MMEJ pathways

Non-homologous end joining (NHEJ) rejoins DNA DSBs as quickly as 30 minutes after break induction with minimal processing [4–6]. Briefly, after a DSB (i.e., DNA ends can be either blunt or possess a short 5′ overhang) has formed, the ring-shaped XRCC6 (X-ray repair cross complementing 6, also known as Ku70)/XRCC5 (X-ray repair cross complementing 5, also known as the Ku80) protein heterodimer quickly binds to the broken DNA ends [4–6]. This binding protects the DNA ends from further resection, preventing MMEJ and HDR pathway initiation [36, 37]. The XRCC6/XRCC5 heterodimer (Ku) then recruits and activates the catalytic subunit of the DNA-dependent protein kinase (DNA-PKcs) [38, 39]. The XRCC6/XRCC5 heterodimer subsequently recruits additional NHEJ factors including XRCC4 (X-ray repair cross complementing 4), NHEJ1 (non-homologous end joining factor 1, also known as XLF), and (DNA ligase IV) to the complex to ligate the DNA DSB ends [40]. Therefore, in the absence of DNA end processing, NHEJ-mediated repair is error-free [40]. In contrast, if the DSB ends are not ligatable due to nucleotide overhangs, DCLRE1C (DNA cross-link repair 1C, also known as Artemis), a single-strand-specific 5′ → 3′ exonuclease, and specialized DNA polymerases POLL (DNA polymerase) and POLM (DNA polymerase μ) generate compatible DNA blunt ends, which can then be ligated by LIG4 [41]. Importantly, this process limits DNA end processing and minimizes mutagenesis (i.e., indels) [41].

3.2 Microhomology-mediated end joining (MMEJ)

Although it was originally thought that most CRISPR/Cas-generated DNA DSBs were repaired by the NHEJ pathway [4–6], it is now apparent that a significant number of these DSBs are also fixed by the MMEJ pathway (>50%) [42, 43]. MMEJ, like NHEJ, does not require a template for repairing DNA DSBs [4–6]. However, in contrast to NHEJ, MMEJ begins with a short-range resection of the DNA DSBs and functions independently of XRCC6/XRCC5 and LIG4 [44]. MMEJ resection is initiated by the MRN (i.e., MRE11 [MRE11 homolog double-strand break repair nuclease]-RAD50 [RAD50 double-strand break repair protein]-NBN [Nibrin, also known as NBS]) DNA DSB repair damage sensing complex with its stimulatory factor RBBP8 (RB-binding protein 8 endonuclease, also known as CTIP) [4–6, 44]. RBBP8 phosphorylation stimulates MRE11 endonuclease activity to create a nick at the 5′ strand near to the DSB, which promotes the removal of XRCC6/XRCC5 and DNA-PKcs, thus preventing NHEJ [45–47]. The resulting nick allows the MRE11 3′-to-5′ exonuclease to resect back toward the DNA DSB, which generates short 3′ overhangs, thus exposing potential single-strand DNA microhomologies (5–25 bps) on opposite strands, which allows the broken ends to realign and anneal [4–6, 44]. Any resulting heterologous 3′ single-stranded DNA (ssDNA) flaps must be removed by the ERCC1 (ERCC excision repair 1 endonuclease non-catalytic subunit)/ERCC4 (ERCC excision repair 4 endonuclease catalytic subunit, also known as XPF) endonuclease [48]. POLQ (DNA polymerase theta) is recruited to stabilize the annealed ssDNA and fills any gaps *via* template-directed DNA synthesis. LigI (DNA ligase 1) or LigIII (DNA ligase 3) subsequently seals the break [49]. Importantly, due to the resection step, MMEJ is "error prone," and therefore, this repair mechanism can lead to indels, chromosomal translocations, and end-to-end chromosomal fusions [42, 43].

3.3 CRISPR/Cas9-induced error-prone end joining DNA repair outcomes

Then, the Cas9/gRNA complex binds to its target site, and the Cas9 HNH nuclease domain cleaves the target strand 3 bp upstream of the PAM site [50]. In contrast, the Cas9 RuvC-like nuclease domain cleaves the non-target strand 3, 4, or 5 bp upstream [50]. Therefore, Cas9-induced DSB ends are either blunt or have 1–2 bp 5′ overhangs [50]. As described above, the blunt ends can be directly ligated with the XRCC4/NHEJ1/LIG4 complex through NHEJ, without any further processing (i.e., "error-free" NHEJ) (**Figure 2**) [42]. Importantly however, even when DNA DSBs are repaired by nontemplated NHEJ, it has been established that the Cas9 cleavage cycle is repeated over and over until NHEJ mutagenic events prevent gRNA target recognition [51]. Thus, this repeated cleavage process enhances the number of non-templated indels (**Figure 2**) [51–55]. Likewise, Cas9-induced DNA DSB ends that have 1–2 bp 5′ overhangs are not ligatable and must be processed further by DCLRE1C, POLL, and POLM, which subsequently generates blunt ends followed by ligation through NHEJ [43]. Importantly, this process results in 1–2 bp indels (**Figure 2**) [55].

Similarly, Cas-induced DSBs repaired by the MMEJ pathway are also innately mutagenic due to the loss of sequence information when the extraneous heterologous 3′ ssDNA flaps are cleaved off [56, 57]. The frequency of deletions mediated by MMEJ is positively correlated with GC base content, and microhomology length, with deletions of two or more nucleotides occurring most often [50–55]. Interestingly, recent studies have established that MMEJ repair outcomes of Cas9-induced DSBs are not random and can be predicted [50–55].

Given the mutagenic nature of Cas9-induced DSBs repaired by NHEJ and MMEJ (i.e., the generation of non-templated indels), this type of end joining is leveraged frequently to silence gene expression (i.e., gene-specific knockout or loss of function)

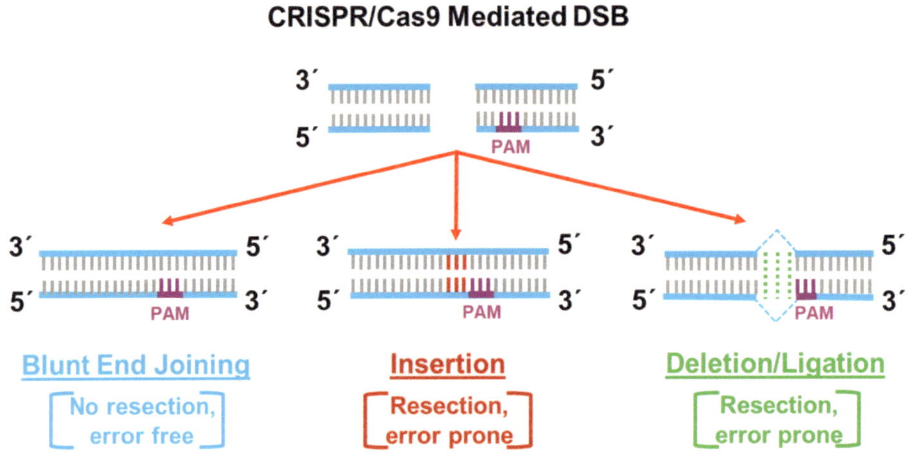

Figure 2.
Schematic representation of potential outcomes of error-prone NHJE/MMEJ repair of CRISPR/Cas9-mediated DSBs. The PAM site (denoted in pink) is shown relative to the DSB generated by CRISPR/Cas9 cleavage. Nontemplated DNA repair is mediated by the NHEJ/MMEJ pathway as described in the text. Three potential outcomes are shown. Nontemplated error-prone repair of CRISPR/Cas9-mediated DSBs can cause frameshift mutations leading to a stop codon and subsequent gene-specific knockout (i.e., loss of function).

(**Figure 2**) [50–54]. Cas9-mediated error-prone NHEJ and MMEJ repair has been utilized to study the function of a wide variety of genes and noncoding elements in cellular and animal models [35, 37, 58]. Additionally, precise template-free end-joining-mediated genome editing through MMEJ has also been achieved [35, 37, 58].

4. Templated homology-directed repair

HDR of endogenously generated DNA DSBs requires extensive DSB end resection and necessitates the physical base pairing interactions between the broken DNA strands and an identical sister chromatid, a homologous chromosome, or an ectopic site (i.e., a double-strand DNA [dsDNA] repair template) [4–6]. Therefore, HDR is most prominent during S and G2 cell cycle phases when an identical sister chromatid is available for recombination [59, 60]. Although HDR is typically an error-free process, indels, point mutations, genomic rearrangements, and subsequent genomic instabilities can result in a DNA donor-dependent or donor-independent manner [61].

4.1 Rad51-dependent homology-directed repair

The repair of DNA DSBs using an endogenous dsDNA repair template can occur through a RAD51 (RAD51 recombinase)-dependent mechanism [32, 34, 35, 62]. Initially, HDR, like MMEJ, begins with a short-range 3′-to-5′ resection (5–25 bps) of DNA ends mediated by the MRN/RBBP8 complex [32, 34, 35, 62]. The short-range resection is then followed by long-range 5′-to-3′ resection (>1000 bps) catalyzed by EXO1 (exonuclease 1) or DNA2 (DNA replication helicase/nuclease 2) with the assistance of BLM (BLM RecQ-like helicase) or WRN (WRN RecQ-like helicase) [32, 34, 35, 62]. The resected 3' ssDNA overhangs are subsequently stabilized by the binding of multiple RPA (heterotrimeric Replication Protein A) complexes [63]. RPA complexes are then replaced by the ATP-dependent nucleoprotein Rad51 (RAD51 recombinase) that forms long helical filaments on the resected 3′ ssDNA overhangs [64, 65]. RAD51 promotes the invasion of the overhangs (i.e., strand exchange), aligns, and pairs the ssDNA with a homologous sister chromatid sequence to form a displacement loop (D-loop) [32, 34, 35, 62]. The invading 3′ ssDNA overhang within the D-loop can then be extended by POLD1 (DNA polymerase delta 1) to synthesize sequences lost at the break site and by end resection using the homologous sister chromatid sequence as a template [66]. Finally, the resulting HDR intermediates can be resolved by multiple mechanisms, which include SDSA (synthesis-dependent strand annealing), crossover and non-crossover dHJ (double Holliday junction), and BIR (break-induced replication) [32, 34, 35, 62].

4.2 Rad51-independent homology-directed repair

Alternatively, endogenously generated DSBs cans also be repaired by a RAD51-independent HDR pathway designated single-strand annealing (SSA). Like Rad51-Dependent Homology-Directed Repair, SSA also requires long-range 5′-to-3′ resection (>1000 bps) catalyzed by EXO1 or DNA2/BLM [32, 34, 35, 62]. The resected 3' ssDNA overhangs are subsequently bound with RPA complexes; however, they are replaced by RAD52 (RAD52 homolog, DNA repair protein), which promotes the annealing of homologous sequences within the two DSB ends [67, 68]. The heterologous DNA flaps generated by SSA annealing are removed by the ERCC1 endonuclease complex, thus producing genomic deletions [48].

4.3 CRISPR/Cas-induced homology-directed repair (HDR) DNA repair outcomes

For HDR, subsequent to CRISPR/Cas-generated DSBs, an exogenous DNA template that shares homology to ends of the DSB and contains the desired gene-specific nucleotide changes, mutations, or additions is required to incorporate these alterations intentionally and precisely *via* the HDR pathway (**Figure 3**) [32, 34, 35]. If a donor DNA template is not provided, then error-prone NHEJ/MMEJ will be the predominant mechanism utilized to repair the DSB and unwanted indels will occur [32, 34, 35].

If exogenous plasmids, PCR products, or chromatinized templates are utilized as dsDNA donor templates, then the Rad51-dependent HDR pathway described above is employed [32, 34, 35, 69]. In contrast, if single-strand oligodeoxynucleotides (ssODNs) are used as homologous donor templates to repair CRISPR/Cas-generated DSBs, then a RAD51-independent mechanism designated, single-stranded DNA donor-templated repair (SSTR) occurs through SSA and synthesis-dependent strand annealing (SDSA) [67, 68, 70, 71]. Like RAD51-dependent HDR, SSTR is initiated by resection of the DSB [67, 72–74] and like SSA, SSTR requires RAD52 to promote annealing of 3′ resected ssDNA tails with ssODN donor templates followed by DNA-templated synthesis [68, 70, 74].

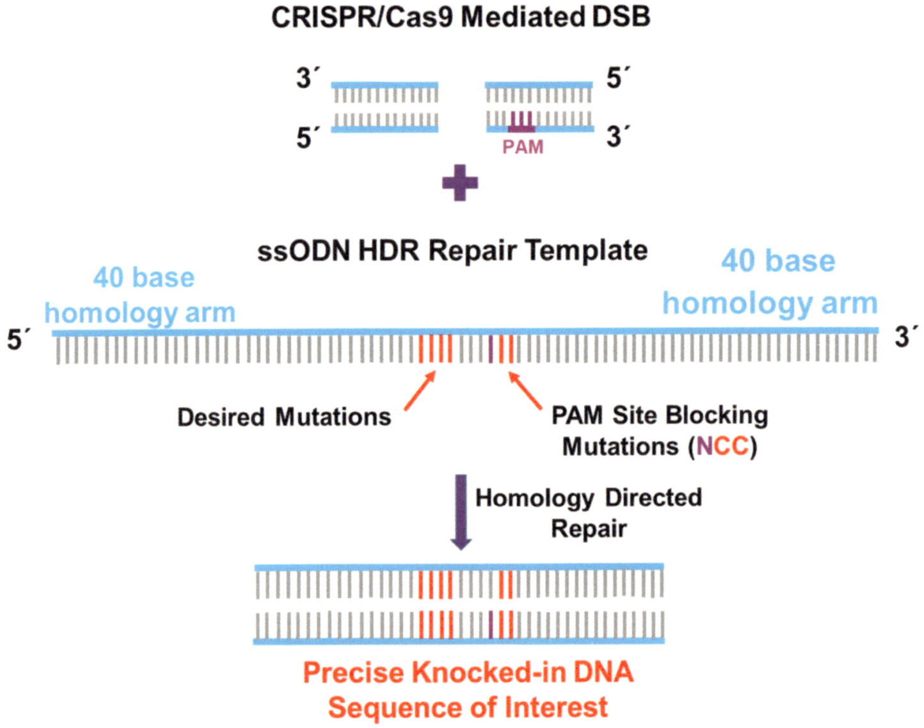

Figure 3.
Schematic representation of precise gene modification mediated by HDR of CRISPR/Cas9-mediated DSBs. The PAM site (denoted in pink) is shown relative to the DSB generated by CRISPR/Cas9 cleavage. The ssODN donor template with symmetric 40-nt homology arms with the desired modifications placed in the middle of the template (denoted four red hash marks). Blocking PAM mutations (i.e., NGG → NCC) are also shown denoted with the pink and two red hash marks). After co-transfection of the Cas/RNP complex with the ssODN donor template, this template is utilized to repair the generated DSB by the HDR pathway. This allows for the precise knock-in of the sequence of interest.

5. Optimizing HDR efficiency

5.1 Allelic considerations

Before initiating any CRISPR/Cas genome editing projects, regardless of whether knockout (i.e., NHEJ/MMEJ) or knock-in (i.e., HDR) experiments are planned, one must explore how many target gene alleles of interest are present in the cell line to be edited. This is a crucial consideration given that many cancer cell lines utilized for gene editing experiments often exhibit extensive somatic gene copy number variation (CNV) [75, 76]. Therefore, a chosen gene of interest could vary from a single copy (e.g., heterozygous deletion), two copies (e.g., normal), several copies (e.g., aneuploidy), or many copies (e.g., gene amplification) depending on which cell line is utilized for the CRISPR/Cas/NHEJ/MMEJ or CRISPR/Cas/HDR experiments.

For example, many CRISPR/Cas studies (i.e., from 2020 to 2022, greater than 100 published as per PubMed) have utilized K562 cells (an immortalized chronic myelogenous leukemia cell line) that are known to contain widespread aneuploidy and numerous obvious structural abnormalities [77–80]. Recently, Zhou et al. [81] published a comprehensive characterization of the K562 genome. This publication proved to be invaluable as our laboratory initiated CRISPR/Cas/HDR studies utilizing an anticancer drug (etoposide)-resistant K562 clonal subline, K/VP.5, previously generated by our laboratory [82, 83].

Briefly, our laboratory studies human DNA topoisomerase IIα (170 kDa, TOP2α/170), which generates transient double-strand DNA breaks to resolve nucleic acid topological entanglements [84, 85]. TOP2α/170 is an important target of anticancer drugs (such as etoposide), whose efficacy is often compromised due to decreased TOP2α/170 levels [86, 87] and resultant attenuation of cytotoxic drug-induced TOP2α-DNA covalent complexes [84, 85]. Compared to parental K562 cells, etoposide-resistant K/VP.5 cells contain reduced TOP2α/170 levels and express high levels of a novel C-terminal truncated TOP2a isoform (90 kDa, TOP2α/90) [88, 89]. TOP2α/90 is the translation product of a short TOP2α mRNA that is generated from a cryptic poly(A) site harbored in intron 19 (i.e., I19 intronic polyadenylation; I19 IPA) [90, 91]. TOP2α/90 lacks the active site tyrosine 805 harbored in exon 20 of full-length TOP2α/170 necessary for TOP2α-mediated DNA strand breaks [88–91]. We hypothesized that, by utilizing CRISPR/Cas/HDR to enhance the TOP2α gene's suboptimal exon 19/intron 19 5′ SS (E19/I19 5′ SS), removal of intron 19 would be enhanced, which in turn would result in decreased TOP2α/90 mRNA/protein, increased TOP2α/170 mRNA/protein, and circumvention of etoposide resistance [92].

Since the human TOP2α gene is harbored on chromosome 17 (i.e., mapped to chromosome 17q21–22) [93], Zhou et al.'s [81] study was utilized to determine the number of TOP2α alleles (i.e., copy number) present in the K562/K/VP.5 cells before initiation of CRISPR/Cas/HDR experiments [92]. It was found that K562 and the isogenic-acquired resistant cell line, K/VP.5, contained three TOP2α alleles [81]. Therefore, our CRISPR/Cas9/HDR strategy was focused on editing all three TOP2α alleles in K/VP.5 cells at the E19/I19 5′ SS to maximize the desired phenotypic change (i.e., decreased TOP2α/90 mRNA/protein, and increased TOP2α/170 mRNA/protein levels) and to circumvent etoposide resistance [92]. qPCR and Sanger sequencing demonstrated that the ratio of wild-type to edited genomic sequence decreased by 1/3 with each allele edited [92]. TOP2α/90 progressively decreased and TOP2α/170 increased with each allele edited by CRISPR/Cas9/HDR. Etoposide resistance was completely reversed when all three TOP2α alleles were edited to enhance the E19/I19

5′ SS [92]. RNA seq confirmed that intron 19 was effectively spliced out in the three allele-edited clone [92].

5.2 PAM site considerations

Multiple studies have demonstrated that CRISPR/Cas-generated DSBs should be in close proximity to the edit site to achieve high HDR efficiencies [93–100]. These investigators established that if a Cas9 PAM site (i.e., NGG; N denotes any nt) was located more than 14 bp (on either DNA strand) from the desired gene-specific nt changes, mutations, or additions, then the efficiency of CRISPR/Cas9/HDR was dramatically reduced. However, Renaud et al. [96] observed that the 14-bp limitation may be pushed to 20 bp utilizing chemically modified ssODN donor templates (see "HDR Considerations" below). Paquet et al. [94] also demonstrated that it was easier to create homozygous gene edits when the PAM site was closer to the intended nucleotide changes and heterozygous gene editing by distance-dependent suboptimal mutation incorporation.

Importantly, Schubert et al. [100] indicated that although guide selection in close proximity with the required HDR changes is important, it was more significant that the gRNA utilized not only targeted Cas9 to the appropriate sequence but also activated Cas9 endonuclease activity. Therefore, since all gRNAs are not equally efficient in activating Cas9, it is essential that the cleavage efficiency for each gRNA utilized is calculated using a T7 endonuclease mismatch cleavage assay (i.e., measuring the extent of indel formation) before initiating HDR experiments [97, 100]. If several Cas9 PAM sites are identified within the 15 base HDR parameter, then the gRNA eliciting the highest cleavage efficiency should be utilized for CRISPR/Cas9/HDR experiments (see "gRNA Considerations" below).

Since the lack of gRNAs with appropriate cleavage efficiency and proximity to the desired HDR-mediated changes is a significant limitation for many CRISPR/Cas9/HDR studies, Schubert et al. [100] also demonstrated that Cas9 D10A nickases (i.e., induce single DNA nicks) can be utilized for HDR mutation experiments if gRNAs target PAM sites on opposite strands of the genomic DNA to generate a staggered DSB provided that the desired mutation is placed between the two nick sites. Alternatively, the number of possible CRISPR/Cas/HDR editing sites can be expanded with the utilization of Cas12a (also known as Cpf1), which recognizes a unique PAM site (TTTV; V denotes an A, C, or G nt) [101].

Since Cas9/Cas12a PAM site recognition restricts targeting and affects CRISPR/Cas/HDR editing efficiency and flexibility, there are efforts to genetically re-engineer CRISPR enzymes to target heretofore inaccessible PAMs [102–105]. For example, Kleinstiver et al. [106] have successfully altered *S. pyogenes* Cas9 (SpCas9) PAM specificity by utilizing bacterial selection-based directed evolution. Walton et al. [104] utilized structure-guided engineering to develop several "near-PAMless" SpCas9 variants capable of targeting NGN and NRN (R denotes an A or G), respectively. Finally, Kleinstiver et al. [105] have also utilized structure-guided protein engineering to improve the targeting range of *Acidaminococcus sp.* Cas12a. Together these studies suggest that the PAM site constraints that currently limit CRISPR/Cas/HDR editing will be circumvented in the future.

5.3 gRNA considerations

Most CRISPR/Cas genome editing experiments are now performed by delivering purified Cas9/Cas12 proteins and chemically synthesized gRNAs as a ribonucleoprotein (RNP) (i.e., Cas/RNP) complex to restrict their temporal activity, improve precision, decrease the immune response, and reduce off-target effects [106, 107]. Specifically, engineered gRNAs have been chemically modified to increase their stability and decrease off-target editing resulting in enhanced cleavage efficiency and improved HDR efficacy [108, 109].

gRNAs can be synthesized in two formats. First, like the endogenous Cas9 gRNA, the crRNA/tracrRNA is a two-piece gRNA where the crRNA (~36–42 nt) and tracrRNA (~67–89 nt) are synthesized as two independent oligonucleotides and are subsequently annealed together through a complimentary linker region to form a functional gRNA [28]. Second, a single guide (sgRNA, 100 nt) can be synthesized, which comprises both the crRNA and tracrRNA in a single oligonucleotide (no annealing is required) (**Figure 4**). It is important to note that the PAM sequence is not included in either gRNA format [28]. One advantage of the two-piece gRNAs is that the tracrRNA sequence is the same for all CRISPR/Cas9 experiments and only the crRNA sequence varies, based on the DNA site to be targeted [28]. Therefore, one chemically synthesized tracrRNA can be annealed to any chemically synthesized crRNA. Chemically synthesized chimeric sgRNAs have the advantage that they exhibit equivalent or greater efficiency compared to the native dual RNA system [109, 110]. We advocate for the use of sgRNAs since our laboratory tested several two-piece gRNAs that exhibited no activity [92] that when resynthesized as sgRNAs displayed high cleavage efficiency [111].

Since RNAs are inherently unstable and susceptible to endo- and exonucleases, considerable effort has been devoted to chemically modifying RNAs to improve their stability. Importantly, Hendel et al. [112] established that chemical modification of

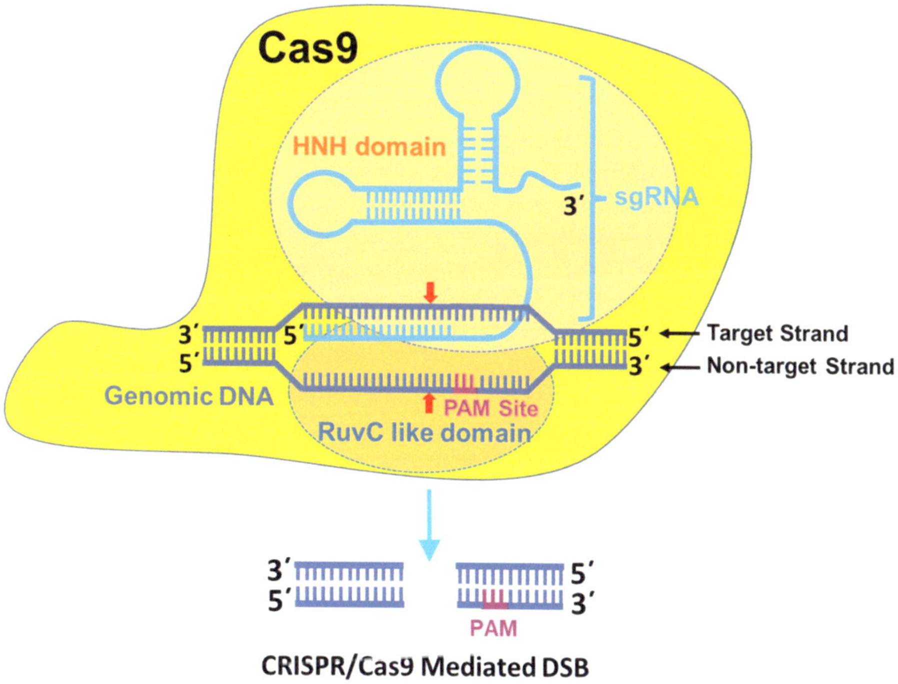

Figure 4.
Schematic representation of the CRISPR/Cas9-mediated DSB with a one-piece sgRNA. The CRISPR/Cas9 schematic is denoted as described in ***Figure 1****, except Cas9 sgRNA is synthesized as a single molecule, which harbors both the crRNA and the tracrRNA (denoted in light blue). The DSB is created as described in* ***Figure 1****. sgRNAs have the advantage that they can exhibit greater efficiency compared to the native dual RNA system with no crRNA/tracrRNA annealing step required [109, 110].*

gRNAs protected them from degradation and enhanced genome editing efficiency. Specifically, these investigators demonstrated that when 2′-O-methyl 3′phosphorothioate (MS), or 2′-O-methyl 3′thioPACE (MSP) chemical modifications were incorporated at both the 5′ and 3′ three terminal nucleotides, and indel formation and HDR were significantly increased [112]. They concluded that chemically synthesized/modified sgRNAs offer significant advantages over sgRNAs expressed by plasmids or by *in vitro* transcription, including 1) scalable and robust production for many applications; 2) greater sgRNA design flexibility; 3) lower toxicity; and 4) increased efficacy [112]. In conclusion, the studies reviewed in this section clearly suggest that the continued optimization of synthetic gRNAs will increase cleavage and on-target efficiency, which will help leading to future efficacious CRISPR-based therapies.

5.4 HDR template considerations

Regardless of whether dsDNA or ssODN donor templates are utilized by distinct HDR pathways to mend Cas RNP complex generated DSBs, the same precise, intentional repair outcomes can be achieved. However, ssODNs donor templates are most frequently used to introduce specific changes (e.g., introduce or correct mutations, and to create short insertions) into specific DNA sequences through HDR due to their superior efficiency, fidelity, and ease of synthesis (**Figure 3**) [71, 74, 95–97, 113]. Importantly, recent studies investigating 1) chemical modifications at the 5′ and 3′ ends of ssODNs donor templates; 2) optimal complementary length; 3) homology arm polarity and asymmetry; and 4) donor template design to prevent the re-cleavage of edited alleles have resulted in empirical rules to rationally design ssODN donor templates to maximize HDR efficiency and flexibility [94–100].

Renaud et al. [96] established that phosphorothioate (PS) chemical modifications at the two terminal nucleotides at both the 5′ and 3′ ends of ssODNs repair templates strongly enhanced genome editing efficiency in cultured cells. These investigators also demonstrated that PS-modified ssODN donor templates also permitted efficient insertion of over 100 nucleotides, while only limited integration was observed with non-chemically ssODNs [96]. Likewise, a higher frequency of insertions was attained in mice and rats using modified ssODNs [96]. The importance of utilizing PS-modified ssODN repair templates to enhance HDR editing efficiency was validated by Liang et al. [97].

Richardson et al. [95] demonstrated that although Cas9 dissociates slowly from dsDNA substrates, Cas9 releases the 3′ end of the cleaved nontarget strand (NT strand, the DNA strand that is not complementary to the gRNA and harbors the "NGG" PAM sequence) before complete Cas9 dissociation. They subsequently showed that ssODNs donor templates complementary to the NT strand increased HDR frequencies compared to donor templates complementary to the target strand (T strand; the DNA strand that is complementary to the gRNA and does not contain the "NGG" PAM sequence) [95]. Finally, these investigators established that ssODN donor templates asymmetrically oriented relative to the 5′- and 3′-side of the generated DSB and complementary to the NT strand also increased HDR rates [95]. In support of these results, Liang et al. [97] also showed that asymmetric ssODNs with 30-bp homology arms 3′ to the insertion and greater than 40 bp of homology at the 5′ end were preferred. This report indicated that the optimal amount of asymmetric ssODN was 10 pmol. However, in contrast to Richardson et al. [95], these investigators only observed a slight increase in HDR efficiency with NT strand compared with T strand ssODN repair templates [97]. Okamoto et al. [99] demonstrated that the optimal ssODN donor template should have a total length of ~75–85 nt with 30 to

35 nt perfectly matched homology arms on the 5′ and 3′ ends and complementary to the gRNA strand (i.e., T strand). Recently, Schubert et al. [100] established further design parameters to improve HDR efficiencies by testing hundreds of genomic loci and multiple cell lines. First, they demonstrated that the ssODN donor template (i.e., NT or T strand) that leads to the highest HDR efficiencies varies greatly depending on the genomic locus and cell type utilized [100]. Second, they observed that the preferred strand (NT or T), relative to the gRNA, is dependent on where the desired HDR modification is located. For example, there is no repair strand preference when the HDR modification is placed precisely at the Cas9 cleavage site [100]. However, if the HDR modifications occur further from the Cas9 cleavage site, a NT strand ssODN donor template is preferred for PAM-distal mutations and a T strand ssODN donor template is ideal for PAM-proximal mutations [100]. Additionally, they showed that asymmetric homology arms did not improve HDR beyond symmetrical homology arms when arm length was ≥30-nt from both the mutation location and the Cas9 cleavage site [100]. These investigators advocated for ssODN donor templates with 40-nt homology arms with modifications placed in the middle (**Figure 3**) [100].

The CAS RNP complex can regenerate DSBs in alleles already appropriately edited, thereby lowering HDR efficiency [94, 99]. In another design innovation using ssODN donor templates, Paquet et al. [94] strategically prevented re-cutting of HDR-edited sites by introducing CRISPR/Cas-blocking PAM site mutations in their repair templates and observed increased HDR accuracy and effectiveness. Okamoto et al. [99] subsequently established that ssODN repair templates with a single mutation in the PAM site (i.e., NGG → NGC) showed the highest HDR efficiency. Their results clearly indicated that the re-cutting of edited alleles resulted in very low HDR efficiencies, and that introducing PAM site mutations within ssODN repair templates to prevent re-cutting is essential for efficient HDR knock-in [99]. Schubert et al. [100] also demonstrated that adding a blocking PAM mutation to the second or third base of the PAM (i.e., NGG → NCG or NGC) in ssODN repair templates resulted in greater HDR efficiency. Donor templates containing two blocking PAM mutations (i.e., NGG → NCC) resulted in the highest HDR efficiency (**Figure 3**). Finally, another important indication for blocking PAM mutations in HDR repair templates is to ensure that when multiple rounds CRISPR/Cas/HDR transfections are required to edit all gene-specific alleles in cell lines that exhibit aneuploidy; the previously edited alleles will not be re-cut in the subsequent rounds of transfection [91, 92].

5.5 Pharmacological strategies to enhance HDR efficiency

Most CRISPR/Cas/HDR genome editing experiments are now performed by transfecting Cas/RNP complexes and ssODN repair templates to restrict temporal activity, thereby reducing off target effects, decreasing immune responses, and increasing HDR efficiency [94–100, 106–110, 112, 113]. Discussion below of pharmacological and genetic strategies for gene editing by HDR will be limited to this experimental paradigm.

Since NHEJ/MMEJ are rapid high-capacity pathways which are active throughout the cell cycle (i.e., G1, S and G2 phases), while HDR is active only after DNA replication is completed and sister chromatids are available to serve as repair templates (i.e., late S and G2 phases) [60], one of the first attempts to pharmacologically enhance CRISPR/Cas/HDR efficiency was to time the delivery of Cas RNA complexes after synchronization of cells using aphidicolin or nocodazole [114]. Lin et al. [114] demonstrated that synchronization, with either aphidicolin or nocodzole, resulted in increased HDR rates (up to 38%) compared with unsynchronized cells.

Since DNA repair is also influenced by the accessibility of DNA binding factors, like Cas RNP complexes [115, 116]. Li et al. [117] hypothesized that CRISPR/Cas/HDR efficiency would be enhanced with histone deacetylase (HDAC) inhibitors, trichostatin A (TSA) and PCI-24781, by promoting a more open chromatin structure [118]. These investigators established that HDR, single strand annealing and ssODN mediated HDR were all increased with HDAC inhibitor treatment [118]. Moreover, this study demonstrated that TSA and PCI-24781 usage also favored HDR by arresting the cell cycle in the G2/M phase [118].

Another pharmacological strategy to improve HDR efficiency is to target the competing NHEJ/MMEJ pathways. Riesenberg et al. [119], explored the efficacy of a wide range of small molecules reported to inhibit the NHEJ/MMEJ pathways or to activate/increase the HDR protein components. These investigators determined that NU7026 (DNA-dependent protein kinase, DNA-PK inhibitor), TSA, MLN4924 (NEDD8 E1 Activating Enzyme Inhibitor), and NSC 15520 (replication protein A1, RPA1 inhibitor) increased HDR efficiencies in various genes and in specified cell lines when DSBs were generated by nickase Cas9n and Cas12/RNP complexes [119]. When Cas9/RNP complexes were utilized to generate DSBs, only NU7026 significantly increased HDR efficacy [119]. NSC 19630 (WRN RecQ like helicase, WRN inhibitor), AICAR (protein kinase AMP-activated catalytic subunit alpha 1, PRKAA1 activator), RS-1 (RAD51 recombinase, RAD51 stimulator), Resveratrol (selective inhibitor of COX-1), SCR7 (DNA ligase IV inhibitor), and L755507 (potent β3-adrenergic receptor partial agonist), showed no clear effect on any Cas/HDR efficiency [119]. Finally, it was demonstrated that the combination of NU7026, TSA, MLN4924, and NSC 15520 resulted in the highest HDR levels observed with Cas9n and Cas12/RNP complexes [119].

In contrast to the results presented above, other investigators have successfully utilized SCR7 (DNA ligase IV inhibitor) to specifically impede the NHEJ pathway to increase HDR activity [120–124]. However, there are also conflicting reports on the ability of this compound to increase HDR [119, 125, 126]. The lack of consistency with this compound regarding HDR efficacy may have resulted from the use of different chemical derivatives of SCR7 [127]. Additionally, Greco et al. [127] demonstrated that SCR7 exhibited greater inhibitory activity against DNA ligases I and III than DNA ligase IV and therefore should target the MMEJ pathway (i.e., also involved in error-prone repair).

6. Conclusion

CRISPR/Cas/HDR is a robust gene editing methodology to purposefully edit out or introduce mutations or insertions at specific genomic sites (i.e., targeted gene knock-in) by creating a DSB along with the introduction of an exogenous template harboring the desired nt changes for DSB repair *via* the HDR pathway [32, 34, 35]. Since DSBs generated in mammalian cells are predominantly repaired by NHEJ/MMEJ pathway [4–6], success of CRISPR/Cas/HDR gene editing will depend on maximizing overall HDR efficacy. We propose the following "workflow" strategies to facilitate the knock-in process (**Figure 5**). First, after developing an experimental CRISPR/Cas/HDR hypothesis, serious consideration must be given to the appropriate cellular system to utilize and the number of gene alleles of interest that may need be edited to obtain a resulting altered phenotype. If the cell line of choice is not well characterized, sequence analysis of CRISPR-edited cells may help determine the number of edited and non-edited alleles [91, 92, 111].

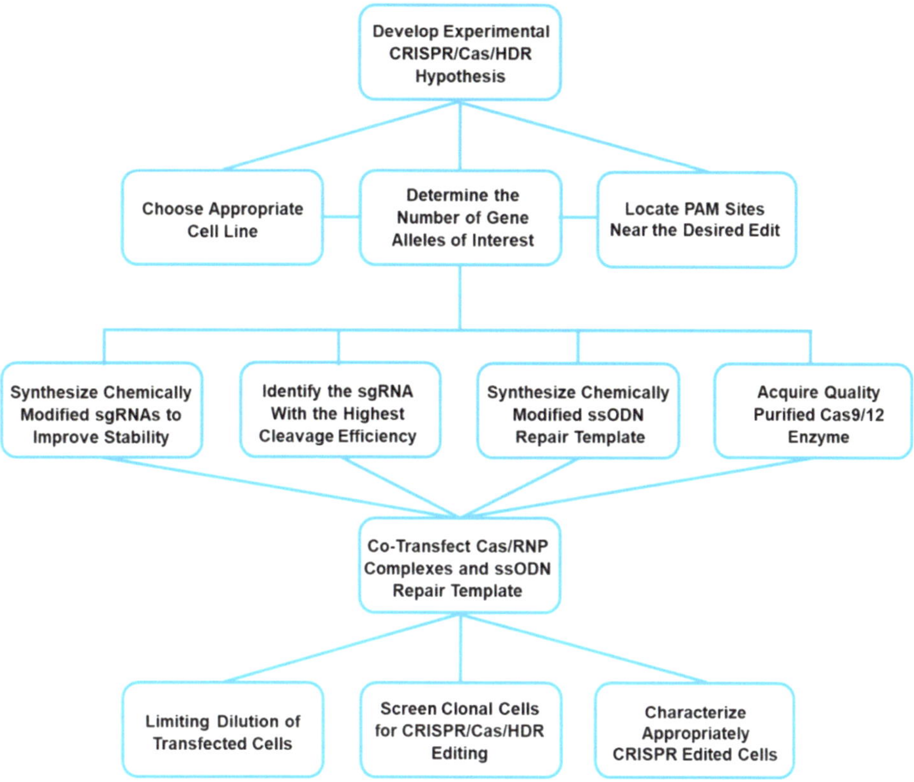

Figure 5.
Proposed experimental flow diagram to maximize CRISPR/Cas/HDR gene editing efficiency. An explanation of the flowchart is discussed in the conclusion section.

Second, the location of the targeted gene knock-in must be analyzed to determine if a Cas9/12 PAM site(s) is/are harbored within 20 nt of the DNA sequence to be edited (both DNA strands should be analyzed) [93–100]. If more than one PAM site is identified that is in close proximity to sequence to be edited, we advocate that multiple chemically modified 5′ and 3′ terminal nucleotide sgRNAs be synthesized. Although chemically modified sgRNAs exhibit increased stability and decreased off-target editing [108, 109], not all gRNAs are equally efficient in activating Cas enzymes [100]. Therefore, the cleavage efficiency for each sgRNA should be determined by employing a T7 endonuclease mismatch cleavage assay [97, 100] before proceeding with CRISPR/Cas/HDR editing experiments. The sgRNA with the highest cleavage efficiency should be utilized [100].

Third, for CRISPR/Cas/HDR, ssODN repair templates chemically modified at their 5′ and 3′ ends should be used to introduce specific changes (e.g., introduce or correct mutations, and to create short insertions) due to their superior efficiency, fidelity, and ease of synthesis [71, 74, 95–97, 113]. Although there are conflicting opinions regarding the optimal homology arm length, homology arm polarity, and asymmetry [94–100], it is now well established that HDR efficiencies varies greatly depending on the genomic locus and cell type utilized [94–100]. The most recent CRISPR/Cas/HDR editing data suggest that ssODN donor templates with symmetric 40-nt homology arms with the desired modifications placed in the middle of the template should be an appropriate standard approach (**Figure 3**) [100]. Regarding the polarity of the ssODN donor templates with respect to the gRNA (see **Figure 1**), an NT strand ssODN donor template is preferred for PAM-distal mutations and a T strand ssODN donor template is ideal for PAM-proximal mutations [100]. Finally, blocking PAM mutations (i.e., NGG → NCC) should always be introduced in ssODN repair templates to further increase HDR efficiency [94, 99, 100] and to allow for potential repeated rounds of CRISPR/Cas/HDR transfections when editing multiple alleles in cell lines, which exhibits aneuploidy (**Figure 3**) [91, 92, 111].

Fourth, the chemically modified sgRNA with the highest cleavage efficiency should be incubated with high-quality, purified Cas9 or Cas12 enzymes that harbor a nuclear localization sequence (NLS) to form a Cas/RNP complex followed by co-transfection with an optimized ssODN donor template as above. Importantly, this procedure improves precision, restricts the temporal activity, decreases the immune response, and reduces the off-target effects of Cas proteins (**Figure 3**) [94–100, 106, 107].

Fifth, after transfection, small aliquots of cell suspension should be sampled to determine genomic cleavage efficiency (i.e., T7 endonuclease mismatch cleavage assay) [97, 100] to validate success in transfection, targeting, and DSB formation by the Cas/RNP complex at the appropriate genome location [91, 111]. Remaining cell suspensions should then be diluted (i.e., limiting dilution) and transferred to 96-well plates at a concentration less than one cell per well and allowed to grow until individual colonies are identifiable in some wells [128]. Individual clonal populations can be split into larger wells and then qPCR and/or Sanger sequencing [91, 92, 111] utilized to determine which clones contain the desired CRISPR/Cas/HDR editing. If electropherogram visualization of genomic sequence reveals only edited sequence, then this clone can be characterized for the hypothesized phenotypic changes. In contrast, if wild-type and edited genomic sequences are identified in the electropherogram, then the ratio of edited to wild-type to edited alleles can be determined. Editing all gene alleles of interest may be required to detect the hypothesized phenotypic change(s) [91, 92, 111]. If sequencing results reveal that a significant number of clonal cells have undergone NHEJ/MMEJ with no HDR editing, then pharmacological inhibitors (as described above) can be considered in an attempt to increase HDR efficiency.

The unique CRISPR/Cas/HDR gene editing experimental outline described in **Figure 5** incorporates the most comprehensive sgRNA and ssODN design considerations along with several important practical details that will help maximize the frequency of precise HDR. It is anticipated that, as strategies to enhance CRISPR/Cas/HDR efficacy continue to advance, that this tractable experimental workflow will accelerate the development of therapeutic gene editing.

Acknowledgements

The work described was supported by grant CA226906-01A1 from the National Institutes of Health (J.C.Y. & T.S.E.).

Author details

Terry S. Elton*, Md. Ismail Hossain, Jessika Carvajal-Moreno, Xinyi Wang, Dalton J. Skaggs and Jack C. Yalowich*
Division of Pharmaceutics and Pharmacology, College of Pharmacy, The Ohio State University, Columbus, Ohio, United States of America

*Address all correspondence to: elton.8@osu.edu and yalowich.1@osu.edu

References

[1] Doudna JA, Charpentier E, Genome editing. The new frontier of genome engineering with CRISPR-Cas9. Science. 2014;**346**:1258096. DOI: 10.1126/science. 1258096

[2] Wright AV, Nuñez JK, Doudna JA. Biology and applications of CRISPR systems: Harnessing Nature's toolbox for Genome engineering. Cell. 2016;**164**:29-44. DOI: 10.1016/j.cell.2015.12.035

[3] Barrangou R, Horvath P. A decade of discovery: CRISPR functions and applications. Nature Microbiology. 2017;**2**:17092. DOI: 10.1038/ nmicrobiol.2017.92

[4] Ceccaldi R, Rondinelli B, D'Andrea AD. Repair pathway choices and consequences at the double-Strand break. Trends in Cell Biology. 2016;**26**:52-64. DOI: 10.1016/j.tcb.2015.07.009

[5] Scully R, Panday A, Elango R, Willis NA. DNA double-strand break repair-pathway choice in somatic mammalian cells. Nature Reviews. Molecular Cell Biology. 2019;**20**:698-714. DOI: 10.1038/s41580-019-0152-0

[6] Zhao B, Rothenberg E, Ramsden DA, Lieber MR. The molecular basis and disease relevance of non-homologous DNA end joining. Nature Reviews. Molecular Cell Biology. 2020;**21**:765-781. DOI: 10.1038/s41580-020-00297-8

[7] Ishino Y, Shinagawa H, Makino K, Amemura M, Nakata A. Nucleotide sequence of the iap gene, responsible for alkaline phosphatase isozyme conversion in Escherichia coli, and identification of the gene product. Journal of Bacteriology. 1987;**169**:5429-5433. DOI: 10.1128/ jb.169.12.5429-5433.1987

[8] Lander ES. The heroes of CRISPR. Cell. 2016;**164**:18-28. DOI: 10.1016/j. cell.2015.12.041

[9] Ishino Y, Krupovic M, Forterre P. History of CRISPR-Cas from encounter with a mysterious repeated sequence to Genome editing technology. Journal of Bacteriology. 2018;**200**:e00580-e00517. DOI: 10.1128/JB.00580-17

[10] Nakata A, Amemura M, Makino K. Unusual nucleotide arrangement with repeated sequences in the Escherichia coli K-12 chromosome. Journal of Bacteriology. 1989;**171**:3553-3556. DOI: 10.1128/jb.171.6.3553-3556.1989

[11] Hermans PW, van Soolingen D, Bik EM, de Haas PE, Dale JW, van Embden JD. Insertion element IS987 from Mycobacterium bovis BCG is located in a hot-spot integration region for insertion elements in mycobacterium tuberculosis complex strains. Infection and Immunity. 1991;**59**:2695-2705. DOI: 10.1128/iai.59.8.2695-2705.1991

[12] Mojica FJ, Juez G, Rodríguez-Valera F. Transcription at different salinities of Haloferax mediterranei sequences adjacent to partially modified PstI sites. Molecular Microbiology. 1993;**9**:613-621. DOI: 10.1111/j.1365-2958.1993.tb01721.x

[13] Mojica FJ, Díez-Villaseñor C, Soria E, Juez G. Biological significance of a family of regularly spaced repeats in the genomes of archaea, bacteria and mitochondria. Molecular Microbiology. 2000;**36**:244-246. DOI: DOI.org/10.1046/ j.1365-2958.2000.01838.x

[14] Jansen R, Embden JD, Gaastra W, Schouls LM. Identification of genes that are associated with DNA repeats in prokaryotes. Molecular Microbiology.

2002;**43**:1565-1575. DOI: 10.1046/j.1365-2958.2002.02839.x

[15] Grissa I, Vergnaud G, Pourcel C. The CRISPRdb database and tools to display CRISPRs and to generate dictionaries of spacers and repeats. BMC Bioinformatics. 2007;**8**:172. DOI: 10.1186/1471-2105-8-172

[16] Haft DH, Selengut J, Mongodin EF, Nelson KE. A guild of 45 CRISPR-associated (Cas) protein families and multiple CRISPR/Cas subtypes exist in prokaryotic genomes. PLoS Computational Biology. 2005;**1**:e60. DOI: 10.1371/journal.pcbi.0010060

[17] Makarova KS, Grishin NV, Shabalina SA, Wolf YI, Koonin EV. A putative RNA-interference-based immune system in prokaryotes: Computational analysis of the predicted enzymatic machinery, functional analogies with eukaryotic RNAi, and hypothetical mechanisms of action. Biology Direct. 2006;**1**:7. DOI: 10.1186/1745-6150-1-7

[18] Makarova KS, Aravind L, Grishin NV, Rogozin IB, Koonin EV. A DNA repair system specific for thermophilic Archaea and bacteria predicted by genomic context analysis. Nucleic Acids Research. 2002;**30**:482-496. DOI: 10.1093/nar/30.2.482

[19] Mojica FJ, Díez-Villaseñor C, García-Martínez J, Soria E. Intervening sequences of regularly spaced prokaryotic repeats derive from foreign genetic elements. Journal of Molecular Evolution. 2005;**60**:174-182. DOI: 10.1007/s00239-004-0046-3

[20] Pourcel C, Salvignol G, Vergnaud G. CRISPR elements in Yersinia pestis acquire new repeats by preferential uptake of bacteriophage DNA and provide additional tools for evolutionary studies. Microbiology. 2005;**151**:653-663. DOI: 10.1099/mic.0.27437-0

[21] Bolotin A, Quinquis B, Sorokin A, Ehrlich SD. Clustered regularly interspaced short palindrome repeats (CRISPRs) have spacers of extrachromosomal origin. Microbiology. 2005;**151**:2551-2561. DOI: 10.1099/mic.0.28048-0

[22] Barrangou R, Fremaux C, Deveau H, Richards M, Boyaval P, Moineau S, et al. CRISPR provides acquired resistance against viruses in prokaryotes. Science. 2007;**315**:1709-1712. DOI: 10.1126/science.1138140

[23] Brouns SJ, Jore MM, Lundgren M, Westra ER, Slijkhuis RJ, Snijders AP, et al. Small CRISPR RNAs guide antiviral defense in prokaryotes. Science. 2008;**321**:960-964. DOI: 10.1126/science.1159689

[24] Mojica FJM, Díez-Villaseñor C, García-Martínez J, Almendros C. Short motif sequences determine the targets of the prokaryotic CRISPR defense system. Microbiology. 2009;**155**:733-740. DOI: 10.1099/mic.0.023960-0

[25] Garneau JE, Dupuis MÈ, Villion M, Romero DA, Barrangou R, Boyaval P, et al. The CRISPR/Cas bacterial immune system cleaves bacteriophage and plasmid DNA. Nature. 2010;**468**:67-71. DOI: 10.1038/nature09523

[26] Sapranauskas R, Gasiunas G, Fremaux C, Barrangou R, Horvath P, Siksnys V. The Streptococcus thermophilus CRISPR/Cas system provides immunity in Escherichia coli. Nucleic Acids Research. 2011;**39**:9275-9282. DOI: 10.1093/nar/gkr606

[27] Deltcheva E, Chylinski K, Sharma CM, Gonzales K, Chao Y, Pirzada ZA, et al. CRISPR RNA maturation by trans-encoded small RNA and host factor RNase III. Nature. 2011;**471**:602-607. DOI: 10.1038/nature09886

[28] Jinek M, Chylinski K, Fonfara I, Hauer M, Doudna JA, Charpentier E. A programmable dual-RNA-guided DNA endonuclease in adaptive bacterial immunity. Science. 2012;**337**:816-821. DOI: 10.1126/science.1225829

[29] Tubbs A, Nussenzweig A. Endogenous DNA damage as a source of genomic instability in cancer. Cell. 2017;**168**:644-656. DOI: 10.1016/j.cell.2017.01.002

[30] Lans H, Hoeijmakers JHJ, Vermeulen W, Marteijn JA. The DNA damage response to transcription stress. Nature Reviews. Molecular Cell Biology. 2019;**20**:766-784. DOI: 10.1038/s41580-019-0169-4

[31] Schumacher B, Pothof J, Vijg J, Hoeijmakers JHJ. The central role of DNA damage in the ageing process. Nature. 2021;**592**:695-703. DOI: 10.1038/s41586-021-03307-7

[32] Yeh CD, Richardson CD, Corn JE. Advances in genome editing through control of DNA repair pathways. Nature Cell Biology. 2019;**21**:1468-1478. DOI: 10.1038/s41556-019-0425-z

[33] Mehta A, Haber JE. Sources of DNA double-strand breaks and models of recombinational DNA repair. Cold Spring Harbor Perspectives in Biology. 2014;**6**:a016428. DOI: 10.1101/cshperspect.a016428

[34] Xue C, Greene EC. DNA repair pathway choices in CRISPR-Cas9-mediated genome editing. Trends in Genetics. 2021;**37**:639-656. DOI: 10.1016/j.tig.2021.02.008

[35] Nambiar TS, Baudrier L, Billon P, Ciccia A. CRISPR-based genome editing through the lens of DNA repair. Molecular Cell. 2022;**82**:348-388. DOI: 10.1016/j.molcel.2021.12.026

[36] Yang SH, Zhou R, Campbell J, Chen J, Ha T, Paull TT. The SOSS1 single-stranded DNA binding complex promotes DNA end resection in concert with Exo1. The EMBO Journal. 2013;**32**:126-139. DOI: 10.1038/emboj.2012.314

[37] Mimitou EP, Symington LS. Ku prevents Exo1 and Sgs1-dependent resection of DNA ends in the absence of a functional MRX complex or Sae2. The EMBO Journal. 2010;**29**:3358-3369. DOI: 10.1038/emboj.2010.193

[38] Davis AJ, Chen BP, Chen DJ. DNA-PK: A dynamic enzyme in a versatile DSB repair pathway. DNA Repair (Amst). 2014 May;**17**:21-29. DOI: 10.1016/j.dnarep.2014.02.020

[39] Frit P, Ropars V, Modesti M, Charbonnier JB, Calsou P. Plugged into the Ku-DNA hub: The NHEJ network. Progress in Biophysics and Molecular Biology. 2019;**147**:62-76. DOI: 10.1016/j.pbiomolbio.2019.03.001

[40] Zhao B, Watanabe G, Morten MJ, Reid DA, Rothenberg E, Lieber MR. The essential elements for the noncovalent association of two DNA ends during NHEJ synapsis. Nature Communications. 2019;**10**:3588. DOI: 10.1038/s41467-019-11507-z

[41] Stinson BM, Moreno AT, Walter JC, Loparo JJ. A mechanism to minimize errors during non-homologous end joining. Molecular Cell. 2020;**77**:1080-1091.e8. DOI: 10.1016/j.molcel.2019.11.018

[42] Bae S, Kweon J, Kim HS, Kim JS. Microhomology-based choice of Cas9 nuclease target sites. Nature Methods. 2014;**11**:705-706. DOI: 10.1038/nmeth.3015

[43] van Overbeek M, Capurso D, Carter MM, Thompson MS, Frias E, Russ C, et al. DNA repair profiling reveals nonrandom outcomes at Cas9-mediated

breaks. Molecular Cell. 2016;**63**:633-646. DOI: 10.1016/j.molcel.2016.06.037

[44] Truong LN, Li Y, Shi LZ, Hwang PY, He J, Wang H, et al. Microhomology-mediated end joining and homologous recombination share the initial end resection step to repair DNA double-strand breaks in mammalian cells. Proceedings of the National Academic Science USA. 2013;**110**:7720-7725. DOI: 10.1073/pnas.1213431110

[45] Chanut P, Britton S, Coates J, Jackson SP, Calsou P. Coordinated nuclease activities counteract Ku at single-ended DNA double-strand breaks. Nature Communications. 2016;**7**:12889. DOI: 10.1038/ncomms12889

[46] Reginato G, Cannavo E, Cejka P. Physiological protein blocks direct the Mre11-Rad50-Xrs2 and Sae2 nuclease complex to initiate DNA end resection. Genes & Development. 2017;**31**:2325-2330. DOI: 10.1101/gad.308254.117

[47] Deshpande RA, Myler LR, Soniat MM, Makharashvili N, Lee L, Lees-Miller SP, et al. DNA-dependent protein kinase promotes DNA end processing by MRN and CtIP. Science Advances. 2020;**6**:eaay0922. DOI: 10.1126/sciadv.aay0922

[48] Ahmad A, Robinson AR, Duensing A, van Drunen E, Beverloo HB, Weisberg DB, et al. ERCC1-XPF endonuclease facilitates DNA double-strand break repair. Molecular and Cellular Biology. 2008;**28**:5082-5092. DOI: 10.1128/MCB.00293-08

[49] Liang L, Deng L, Nguyen SC, Zhao X, Maulion CD, Shao C, et al. Human DNA ligases I and III, but not ligase IV, are required for microhomology-mediated end joining of DNA double-strand breaks. Nucleic Acids Research. 2008;**36**:3297-3310. DOI: 10.1093/nar/gkn184

[50] Shou J, Li J, Liu Y, Wu Q. Precise and predictable CRISPR chromosomal rearrangements reveal principles of Cas9-mediated nucleotide insertion. Molecular Cell. 2018;**71**:498-509.e4. DOI: 10.1016/j.molcel.2018.06.021

[51] Brinkman EK, Chen T, de Haas M, Holland HA, Akhtar W, van Steensel B. Kinetics and Fidelity of the repair of Cas9-induced double-Strand DNA breaks. Molecular Cell. 2018;**70**:801-813.e6. DOI: 10.1016/j.molcel.2018.04.016

[52] Allen F, Crepaldi L, Alsinet C, Strong AJ, Kleshchevnikov V, De Angeli P, et al. Predicting the mutations generated by repair of Cas9-induced double-strand breaks. Nature Biotechnology. 2018;**37**:64-72. DOI: 10.1038/nbt.4317

[53] Chakrabarti AM, Henser-Brownhill T, Monserrat J, Poetsch AR, Luscombe NM, Scaffidi P. Target-specific precision of CRISPR-mediated Genome editing. Molecular Cell. 2019;**73**:699-713. DOI: 10.1016/j.molcel.2018.11.031

[54] Chen W, McKenna A, Schreiber J, Haeussler M, Yin Y, Agarwal V, et al. Massively parallel profiling and predictive modeling of the outcomes of CRISPR/Cas9-mediated double-strand break repair. Nucleic Acids Research. 2019;**47**:7989-8003. DOI: 10.1093/nar/gkz487

[55] Shen MW, Arbab M, Hsu JY, Worstell D, Culbertson SJ, Krabbe O, et al. Predictable and precise template-free CRISPR editing of pathogenic variants. Nature. 2018;**563**:646-651. DOI: 10.1038/s41586-018-0686-x

[56] Sfeir A, Symington LS. Microhomology-mediated end joining: A Back-up survival mechanism or dedicated pathway? Trends in

Biochemical Sciences. 2015;**40**:701-714. DOI: 10.1016/j.tibs.2015.08.006

[57] Seol JH, Shim EY, Lee SE. Microhomology-mediated end joining: Good, bad and ugly. Mutation Research. 2018;**809**:81-87. DOI: 10.1016/j.mrfmmm.2017.07.002

[58] Pickar-Oliver A, Gersbach CA. The next generation of CRISPR-Cas technologies and applications. Nature Reviews. Molecular Cell Biology. 2019;**20**:490-507. DOI: 10.1038/s41580-019-0131-5

[59] San Filippo J, Sung P, Klein H. Mechanism of eukaryotic homologous recombination. Annual Review of Biochemistry. 2008;**77**:229-257. DOI: 10.1146/annurev.biochem.77.061306.125255

[60] Karanam K, Kafri R, Loewer A, Lahav G. Quantitative live cell imaging reveals a gradual shift between DNA repair mechanisms and a maximal use of HR in mid S phase. Molecular Cell. 2012;**47**:320-329. DOI: 10.1016/j.molcel.2012.05.052

[61] Al-Zain AM, Symington LS. The dark side of homology-directed repair. DNA Repair (Amst). 2021;**106**:103181. DOI: 10.1016/j.dnarep.2021.103181

[62] Cejka P, Symington LS. DNA end resection: Mechanism and control. Annual Review of Genetics. 2021;**55**:285-307. DOI: 10.1146/annurev-genet-071719-020312

[63] Chen R, Wold MS. Replication protein A: Single-stranded DNA's first responder: Dynamic DNA-interactions allow replication protein A to direct single-strand DNA intermediates into different pathways for synthesis or repair. BioEssays. 2014;**36**:1156-1161. DOI: 10.1002/bies.201400107

[64] Shrivastav M, De Haro LP, Nickoloff JA. Regulation of DNA double-strand break repair pathway choice. Cell Research. 2008;**18**:134-147. DOI: 10.1038/cr.2007.111

[65] Xu J, Zhao L, Xu Y, Zhao W, Sung P, Wang HW. Cryo-EM structures of human RAD51 recombinase filaments during catalysis of DNA-strand exchange. Nature Structural & Molecular Biology. 2017;**24**:40-46. DOI: 10.1038/nsmb.3336

[66] McVey M, Khodaverdian VY, Meyer D, Cerqueira PG, Heyer WD. Eukaryotic DNA polymerases in homologous recombination. Annual Review of Genetics. 2016;**50**:393-421. DOI: 10.1146/annurev-genet-120215-035243

[67] Gallagher DN, Haber JE. Repair of a site-specific DNA cleavage: Old-school lessons for Cas9-mediated gene editing. ACS Chemical Biology. 2018;**13**:397-405. DOI: 10.1021/acschembio.7b00760

[68] Gallagher DN, Pham N, Tsai AM, Janto NV, Choi J, Ira G. Haber JE.A Rad51-independent pathway promotes single-strand template repair in gene editing. PLoS Genetics. 2020;**16**:e1008689. DOI: 10.1371/journal.pgen.1008689

[69] Cruz-Becerra G, Kadonaga JT. Enhancement of homology-directed repair with chromatin donor templates in cells. eLife. 2020;**9**:e55780. DOI: 10.7554/eLife.55780

[70] Storici F, Snipe JR, Chan GK, Gordenin DA, Resnick MA. Conservative repair of a chromosomal double-strand break by single-strand DNA through two steps of annealing. Molecular and Cellular Biology. 2006;**26**:7645-7657. DOI: 10.1128/MCB.00672-06

[71] Richardson CD, Kazane KR, Feng SJ, Zelin E, Bray NL, Schäfer AJ,

et al. CRISPR-Cas9 genome editing in human cells occurs via the Fanconi anemia pathway. Nature Genetics. 2018;**50**:1132-1139. DOI: 10.1038/s41588-018-0174-0

[72] Canny MD, Moatti N, Wan LCK, Fradet-Turcotte A, Krasner D, Mateos-Gomez PA, et al. Inhibition of 53BP1 favors homology-dependent DNA repair and increases CRISPR-Cas9 genome-editing efficiency. Nature Biotechnology. 2018;**36**:95-102. DOI: 10.1038/nbt.4021

[73] Nambiar TS, Billon P, Diedenhofen G, Hayward SB, Taglialatela A, Cai K, et al. Stimulation of CRISPR-mediated homology-directed repair by an engineered RAD18 variant. Nature Communications. 2019;**10**:3395. DOI: 10.1038/s41467-019-11105-z

[74] Gallagher DN, Haber JE. Single-strand template repair: Key insights to increase the efficiency of gene editing. Current Genetics. 2021;**67**:747-753. DOI: 10.1007/s00294-021-01186-z

[75] Beroukhim R, Mermel CH, Porter D, Wei G, Raychaudhuri S, Donovan J, et al. The landscape of somatic copy-number alteration across human cancers. Nature. 2010;**463**:899-905. DOI: 10.1038/nature08822

[76] Santaguida S, Amon A. Short- and long-term effects of chromosome mis-segregation and aneuploidy. Nature Reviews. Molecular Cell Biology. 2015;**16**:473-485. DOI: 10.1038/nrm4025

[77] Selden JR, Emanuel BS, Wang E, Cannizzaro L, Palumbo A, Erikson J, et al. Amplified Cλ and c-abl genes are on the same marker chromosome in K562 leukemia cells. Proceedings of the National Academy of Sciences. 1983;**80**:7289-7292. DOI: 10.1073/pnas.80.23.7289

[78] Wu SQ, Voelkerding KV, Sabatini L, Chen XR, Huang J, Meisner LF. Extensive amplification of bcr/abl fusion genes clustered on three marker chromosomes in human leukemic cell line K-562. Leukemia. 1995;**9**:858-862

[79] Gribble SM, Roberts I, Grace C, Andrews KM, Green AR, Nacheva EP. Cytogenetics of the chronic myeloid leukemia-derived cell line K562. Cancer Genetics and Cytogenetics. 2000;**118**:1-8. DOI: 10.1016/S0165-4608(99)00169-7

[80] Naumann S, Reutzel D, Speicher M, Decker HJ. Complete karyotype characterization of the K562 cell line by combined application of G-banding, multiplex-fluorescence in situ hybridization, fluorescence in situ hybridization, and comparative genomic hybridization. Leukemia Research. 2001;**25**:313-322. DOI: 10.1016/S0145-2126(00)00125-9

[81] Zhou B, Ho SS, Greer SU, Zhu X, Bell JM, Arthur JG, et al. Comprehensive, integrated, and phased whole-genome analysis of the primary ENCODE cell line K562. Genome Research. 2019;**29**:472-484. DOI: 10.1101/gr.234948.118

[82] Ritke MK, Yalowich JC. Altered gene expression in human leukemia K562 cells selected for resistance to etoposide. Biochemical Pharmacology. 1993;**46**:2007-2020. DOI: 10.1016/0006-2952(93)90643-b

[83] Ritke MK, Roberts D, Allan WP, Raymond J, Bergoltz VV, Yalowich JC. Altered stability of etoposide-induced topoisomerase II-DNA complexes in resistant human leukaemia K562 cells. British Journal of Cancer. 1994;**69**:687-697. DOI: 10.1038/bjc.1994.131

[84] Deweese JE, Osheroff N. The DNA cleavage reaction of topoisomerase II: wolf in sheep's clothing. Nucleic Acids

Research. 2009;**37**:738-748. DOI: 10.1093/nar/gkn937

[85] Nitiss JL. Targeting DNA topoisomerase II in cancer chemotherapy. Nature Reviews. Cancer. 2009;**9**:338-350. DOI: 10.1038/nrc2607

[86] Ganapathi RN, Ganapathi MK. Mechanisms regulating resistance to inhibitors of topoisomerase II. Frontiers in Pharmacology. 2013;**4**:89. DOI: 10.3389/fphar.2013.00089

[87] Capelôa T, Benyahia Z, Zampieri LX, Blackman MCNM, Sonveaux P. Metabolic and non-metabolic pathways that control cancer resistance to anthracyclines. Seminars in Cell & Developmental Biology. 2020;**98**:181-191. DOI: 10.1016/j.semcdb.2019.05.006

[88] Kanagasabai R, Serdar L, Karmahapatra S, Kientz CA, Ellis J, Ritke MK, et al. Alternative RNA processing of topoisomerase IIα in etoposide-resistant human leukemia K562 cells: Intron retention results in a novel C-terminal truncated 90-kDa isoform. The Journal of Pharmacology and Experimental Therapeutics. 2017;**360**:152-163

[89] Kanagasabai R, Karmahapatra S, Kientz CA, Yu Y, Hernandez VA, Kania EE, et al. The novel C-terminal truncated 90-kDa isoform of topoisomerase IIα (TOP2α/90) is a determinant of etoposide resistance in K562 leukemia cells via heterodimerization with the TOP2α/170 isoform. Molecular Pharmacology. 2018;**93**:515-525. DOI: 10.1124/mol.117.111567

[90] Elton TS, Ozer HG, Yalowich JC. Effects of DNA topoisomerase IIα splice variants on acquired drug resistance. Cancer Drug Resistance. 2020;**3**:161-170

[91] Elton TS, Hernandez VA, Carvajal-Moreno J, Wang X, Ipinmoroti D, Yalowich JC. Intronic polyadenylation in acquired cancer drug resistance circumvented by utilizing CRISPR/Cas9 with homology-directed repair: The tale of human DNA topoisomerase IIα. Cancers (Basel). 2022;**14**:3148. DOI: 10.3390/cancers14133148

[92] Hernandez VA, Carvajal-Moreno J, Papa JL, Shkolnikov N, Li J, Ozer HG, et al. CRISPR/Cas9 Genome editing of the human topoisomerase IIα intron 19 5' splice site circumvents etoposide resistance in human Leukemia K562 cells. Molecular Pharmacology. 2021;**99**:226-241. DOI: 10.1124/molpharm.120.000173

[93] Lang AJ, Mirski SE, Cummings HJ, Yu Q, Gerlach JH, Cole SP. Structural organization of the human TOP2A and TOP2B genes. Gene. 1998;**221**:255-266. DOI: 10.1016/s0378-1119(98)00468-5

[94] Paquet D, Kwart D, Chen A, Sproul A, Jacob S, Teo S, et al. Efficient introduction of specific homozygous and heterozygous mutations using CRISPR/Cas9. Nature. 2016;**533**:125-129. DOI: 10.1038/nature17664

[95] Richardson CD, Ray GJ, DeWitt MA, Curie GL, Corn JE. Enhancing homology-directed genome editing by catalytically active and inactive CRISPR-Cas9 using asymmetric donor DNA. Nature Biotechnology. 2016;**34**:339-344. DOI: 10.1038/nbt.3481

[96] Renaud JB, Boix C, Charpentier M, De Cian A, Cochennec J, Duvernois-Berthet E, et al. Improved Genome editing efficiency and flexibility using modified oligonucleotides with TALEN and CRISPR-Cas9 nucleases. Cell Reports. 2016;**14**:2263-2272. DOI: 10.1016/j.celrep.2016.02.018

[97] Liang X, Potter J, Kumar S, Ravinder N, Chesnut JD. Enhanced CRISPR/Cas9-mediated precise genome editing by

improved design and delivery of gRNA, Cas9 nuclease, and donor DNA. Journal of Biotechnology. 2017;**241**:136-146. DOI: 10.1016/j.jbiotec.2016.11.011

[98] O'Brien AR, Wilson LOW, Burgio G, Bauer DC. Unlocking HDR-mediated nucleotide editing by identifying high-efficiency target sites using machine learning. Scientific Reports. 2019;**9**:2788. DOI: 10.1038/s41598-019-39142-0

[99] Okamoto S, Amaishi Y, Maki I, Enoki T, Mineno J. Highly efficient genome editing for single-base substitutions using optimized ssODNs with Cas9-RNPs. Scientific Reports. 2019;**9**:4811. DOI: 10.1038/s41598-019-41121-4

[100] Schubert MS, Thommandru B, Woodley J, Turk R, Yan S, Kurgan G, et al. Optimized design parameters for CRISPR Cas9 and Cas12a homology-directed repair. Scientific Reports. 2021;**11**:19482. DOI: 10.1038/s41598-021-98965-y

[101] Zetsche B, Gootenberg JS, AbudayyehOO,SlaymakerIM,MakarovaKS, Essletzbichler P, et al. Cpf1 is a single RNA-guided endonuclease of a class 2 CRISPR-Cas system. Cell. 2015;**163**:759-771. DOI: 10.1016/j.cell.2015.09.038

[102] Anders C, Niewoehner O, Duerst A, Jinek M. Structural basis of PAM-dependent target DNA recognition by the Cas9 endonuclease. Nature. 2014;**513**:569-573. DOI: 10.1038/nature13579

[103] Kleinstiver BP, Prew MS, Tsai SQ, Topkar VV, Nguyen NT, Zheng Z, et al. Engineered CRISPR-Cas9 nucleases with altered PAM specificities. Nature. 2015;**523**:481-485. DOI: 10.1038/nature14592

[104] Walton RT, Christie KA, Whittaker MN, Kleinstiver BP. Unconstrained genome targeting with near-PAMless engineered CRISPR-Cas9 variants. Science. 2020;**368**:290-296. DOI: 10.1126/science.aba8853

[105] Kleinstiver BP, Sousa AA, Walton RT, Tak YE, Hsu JY, Clement K, et al. Engineered CRISPR-Cas12a variants with increased activities and improved targeting ranges for gene, epigenetic and base editing. Nature Biotechnology. 2019;**37**:276-282. DOI: 10.1038/s41587-018-0011-0

[106] Sun L, Wu J, Du F, Chen X, Chen ZJ. Cyclic GMP-AMP synthase is a cytosolic DNA sensor that activates the type I interferon pathway. Science. 2013;**339**:786-791. DOI: 10.1126/science.1232458

[107] Kim S, Kim D, Cho SW, Kim J, Kim JS. Highly efficient RNA-guided genome editing in human cells via delivery of purified Cas9 ribonucleoproteins. Genome Research. 2014;**24**:1012-1019. DOI: 10.1101/gr.171322.113

[108] Allen D, Rosenberg M, Hendel A. Using synthetically engineered guide RNAs to enhance CRISPR Genome editing systems in mammalian cells. Frontiers in Genome Edition. 2021;**2**:617910. DOI: 10.3389/fgeed.2020.617910

[109] Kelley ML, Strezoska Ž, He K, Vermeulen A, Av S. Versatility of chemically synthesized guide RNAs for CRISPR-Cas9 genome editing. Journal of Biotechnology. 2016;**233**:74-83. DOI: 10.1016/j.jbiotec.2016.06.01

[110] Shapiro J, Iancu O, Jacobi AM, McNeill MS, Turk R, Rettig GR, et al. Increasing CRISPR efficiency and measuring its specificity in HSPCs using a clinically relevant system. Molecular Therapy in Methods Clinical Development. 2020;**17**:1097-1107. DOI: 10.1016/j.omtm.2020.04.027

[111] Hernandez VA, Carvajal-Moreno J, Wang X, Pietrzak M, Yalowich JC,

Elton TS. Use of CRISPR/Cas9 with homology-directed repair to silence the human topoisomerase IIα intron-19 5′ splice site: Generation of etoposide resistance in human leukemia K562 cells. PLoS One. 2022;**17**:e0265794. DOI: 10.1371/journal.pone.0265794

[112] Hendel A, Bak RO, Clark JT, Kennedy AB, Ryan DE, Roy S, et al. Chemically modified guide RNAs enhance CRISPR-Cas genome editing in human primary cells. Nature Biotechnology. 2015;**33**:985-989. DOI: 10.1038/nbt.3290

[113] Chen F, Pruett-Miller SM, Huang Y, Gjoka M, Duda K, Taunton J, et al. High-frequency genome editing using ssDNA oligonucleotides with zinc-finger nucleases. Nature Methods. 2011;**8**:753-755. DOI: 10.1038/nmeth.1653

[114] Lin S, Staahl BT, Alla RK, Doudna JA. Enhanced homology-directed human genome engineering by controlled timing of CRISPR/Cas9 delivery. eLife. 2014;**15**(3):e04766. DOI: 10.7554/eLife.04766

[115] Chen X, Rinsma M, Janssen JM, Liu J, Maggio I, Gonçalves MA. Probing the impact of chromatin conformation on genome editing tools. Nucleic Acids Research. 2016;**44**:6482-6492. DOI: 10.1093/nar/gkw524

[116] Chen X, Liu J, Janssen JM, Gonçalves MAFV. The chromatin structure differentially impacts high-specificity CRISPR-Cas9 nuclease strategies. Molecular Therapy in Nucleic Acids. 2017;**8**:558-563. DOI: 10.1016/j.omtn.2017.08.005

[117] Li G, Zhang X, Wang H, Liu D, Li Z, Wu Z, et al. Increasing CRISPR/Cas9-mediated homology-directed DNA repair by histone deacetylase inhibitors. The International Journal of Biochemistry & Cell Biology. 2020;**125**:105790. DOI: 10.1016/j.biocel.2020.105790

[118] Eberharter A, Becker PB. Histone acetylation: A switch between repressive and permissive chromatin. Second in review series on chromatin dynamics. EMBO Reports. 2002;**3**:224-229. DOI: 10.1093/embo-reports/kvf053

[119] Riesenberg S, Maricic T. Targeting repair pathways with small molecules increases precise genome editing in pluripotent stem cells. Nature Communications. 2018;**9**:2164. DOI: 10.1038/s41467-018-04609-7

[120] Maruyama T, Dougan SK, Truttmann MC, Bilate AM, Ingram JR, Ploegh HL. Increasing the efficiency of precise genome editing with CRISPR-Cas9 by inhibition of nonhomologous end joining. Nature Biotechnology. 2015;**33**:538-542. DOI: 10.1038/nbt.3190

[121] Chu VT, Weber T, Wefers B, Wurst W, Sander S, Rajewsky K, et al. Increasing the efficiency of homology-directed repair for CRISPR-Cas9-induced precise gene editing in mammalian cells. Nature Biotechnology. 2015;**33**:543-548. DOI: 10.1038/nbt.3198

[122] Pinder J, Salsman J, Dellaire G. Nuclear domain 'knock-in' screen for the evaluation and identification of small molecule enhancers of CRISPR-based genome editing. Nucleic Acids Research. 2015;**43**:9379-9392. DOI: 10.1093/nar/gkv993

[123] Robert F, Barbeau M, Éthier S, Dostie J, Pelletier J. Pharmacological inhibition of DNA-PK stimulates Cas9-mediated genome editing. Genome Medicine. 2015;**7**:93. DOI: 10.1186/s13073-015-0215-6

[124] Hu Z, Shi Z, Guo X, Jiang B, Wang G, Luo D, et al. Ligase IV inhibitor

SCR7 enhances gene editing directed by CRISPR-Cas9 and ssODN in human cancer cells. Cell & Bioscience. 2018;**8**:12. DOI: 10.1186/s13578-018-0200-z

[125] Gutschner T, Haemmerle M, Genovese G, Draetta GF, Chin L. Post-translational regulation of Cas9 during G1 enhances homology-directed repair. Cell Reports. 2016;**14**:1555-1566. DOI: 10.1016/j.celrep.2016.01.019

[126] Zhang JP, Li XL, Li GH, Chen W, Arakaki C, Botimer GD, et al. Efficient precise knockin with a double cut HDR donor after CRISPR/Cas9-mediated double-stranded DNA cleavage. Genome Biology. 2017;**18**:35. DOI: 10.1186/s13059-017-1164-8

[127] Greco GE, Matsumoto Y, Brooks RC, Lu Z, Lieber MR, Tomkinson AE. SCR7 is neither a selective nor a potent inhibitor of human DNA ligase IV. DNA Repair (Amst). 2016;**43**:18-23. DOI: 10.1016/j.dnarep.2016.04.004

[128] Hu P, Zhang W, Xin H, Deng G. Single cell isolation and analysis. Frontiers in Cell and Development Biology. 2016;**4**:116. DOI: 10.3389/fcell.2016.00116

Chapter 4

The Prominent Characteristics of the Effective sgRNA for a Precise CRISPR Genome Editing

Reza Mohammadhassan, Sara Tutunchi, Negar Nasehi, Fatemeh Goudarziasl and Lena Mahya

Abstract

Clustered regularly interspaced short palindromic repeats (CRISPRs) technique is the most effective and novelist technique for genome editing. CRISPR mechanism has been widely developed for gene editing, gene silencing, high-specific regulation of the transcription, and reducing off-target effects through double-strand breaks (DSBs) in the genomic DNA and then modifying nucleotide sequences of the target gene in diverse plant and animal species. However, the application may be restricted by a high rate of off-target effects. So, there are many studies on designing precise single-guide RNAs (sgRNAs) to minimize off-target effects. Thus, the high-efficiency design of a specific sgRNA is critical. First, in the chapter, the sgRNA origin and different types of gRNA will be outlined. Then, the off-target effect will be described. Next, the remarkable characteristics of the sgRNA will be highlighted to improve precise gene editing. Finally, some popular *in silico* tools will be introduced for designing sgRNA.

Keywords: sgRNA, guide RNA, crRNA, tracrRNA, off-target effect, designing tools, CRISPR/Cas

1. Introduction

Clustered regularly interspaced short palindromic repeats (CRISPRs) and their CRISPR-associated (Cas) proteins system is an effective immune system among bacteria and archaea. This system was first discovered in the *E.coli* genome. CRISPR/Cas is an acquired immunity mechanism in many bacteria and archaea against the genome of the infection factors such as viruses and plasmids [1]. CRISPR/Cas system is classified into three major groups (I, II, and III) with a specific functional mechanism and gene family encoding the specific Cas proteins. Types I and III apply several Cas proteins for endonuclease activity, while type II uses only one protein (Cas9) [2]. Evolutionary competition between the pathogens and host in the CRISPR/Cas system shows a very high variable rate in structures and functions. So, recent classification has stated that the CRISPR system has been categorized into two classes (I and II) and six types (I–VI) [3].

 IntechOpen

Most studies on genome engineering have been performed in system type II, derived from Streptococcus pyogenes (SpCas9). The advantage of system type II is that it needs only one protein (Cas9) for endonuclease activity. However, this system also needs the types of RNA, including CRISPR RNA (crRNA), which functions as a Cas protein guide for pairing with the target genome sequences, and trans-activating CRISPR RNA (tracrRNA), which play critical roles in crRNA maturation and directing Cas9 to the desired site [4, 5]. Merging of tracrRNA:crRNA sequences as a chimeric sequence known as single-guide RNA (sgRNA), which covers the features of both RNA types, makes it suitable and applicable for genome editing [6].

All CRISPR sites cover consecutive and spacer repetitions. These consecutive repetitions include identical sequences, while the spacer sequences originate from the genome of foreign factors [7, 8]. CRISPR sites and Cas proteins develop acquired immunity against the invading DNA. Suppose a microorganism survives the invasion of a pathogen. In that case, the integrated CRISPR system will be able to incorporate a piece of the pathogen DNA into its genome and then use it to fight against subsequent infections. This bacterial immune system degenerates the phage genome by integrating short fragments of the phage DNA in the spacer region of the CRISPR sites and transcribing the spacers (known as crRNAs) with associated Cas endonuclease in subsequent infections [9–11].

Briefly, the CRISPR system, as the RNA-mediated immune system in prokaryotes (bacteria and archaea), functions in four stages (**Figure 1**):

1. *Admission*: short viral or plasmid DNA fragments are recognized and then inserted as a spacer between two consecutive repetitions into the CRISPR sites [12, 13].

2. *Expression*: the CRISPR sites are transcribed to a long precursor crRNA (pre-crRNA) containing a complete array of CRISPR repetitions and sequences derived from infectious factors [14].

3. *Insertion*: Pre-crRNA is cleaved by a special endoribonuclease into small guide sequences known as crRNA [15].

4. *Targeting*: The crRNAs guide Cas endonuclease to cleave complementary DNA or RNA sequences flanked by a protospacer adjacent motif (PAM) (for DNA targets) or a protospacer flanking sequence (usually known as PFS) without significant complement to the crRNA repeats (for RNA targets) [16, 17].

The classification represents the evolution of subtype-specific molecular defensive mechanisms for crRNA expression and maturation, as well as the inhibition of infectious factors [18]. The main known vital component of the CRISPR/Cas system is crRNA, which is common in types I and III. Pre-crRNAs are initially cleaved within the repeats by a Cas6 endoribonucleases family, and then intermediate crRNAs are further matured to generate shorter repeat-spacer crRNAs in type III. In both type I and type III, mature crRNAs direct a complex of multiple Cas proteins to the cognate-invading nucleic acids. Then, the target nucleic acids are cleaved by a Cas endonuclease of the ribonucleoprotein complex [19–21].

Pre-crRNA processing necessitates base-pairing of each pre-crRNA repeat with tracrRNA, a small noncoding RNA encoded near the *Cas* genes and spacer array [7].

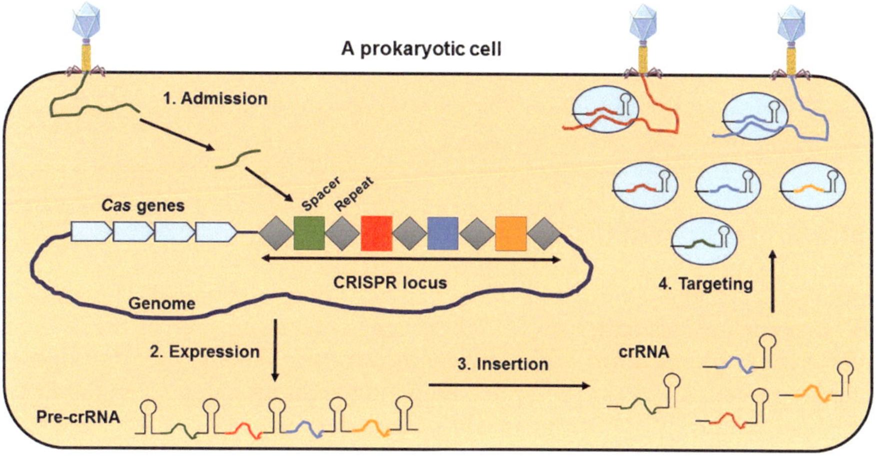

Figure 1.
RNA-mediated CRISPR immune system in.

The base-pairing drives cleavage and binding by RNase III and Cas9, respectively. Then, the crRNA:tracrRNA complex can direct Cas9 to bind target DNA sequences by matching PAM [16, 17]. In addition, type II CRISPR-Cas systems continuously utilize the crRNA:tracrRNA complex to identify and cleave double-stranded DNA (dsDNA). Recognition is driven by base-pairing between the guide sequence and the RNA in the crRNA:tracrRNA complex (**Figure 2**) [22, 23]. The tracrRNA is a common component of CRISPR-Cas systems among all three subtypes of type II systems and is required for crRNA biosynthesis [24]. These basic principles were discovered by following efforts to characterize tracrRNAs and crRNA biogenesis [25]. The biogenesis of crRNA, the structures of the crRNA:tracrRNA complex, and tracrRNA genomic location are variable among these subtypes [26]. The tracrRNA discovery as a key factor of crRNA biosynthesis allowed the sgRNA to be invented and Cas9 to be adopted as the core component of CRISPR technology [27].

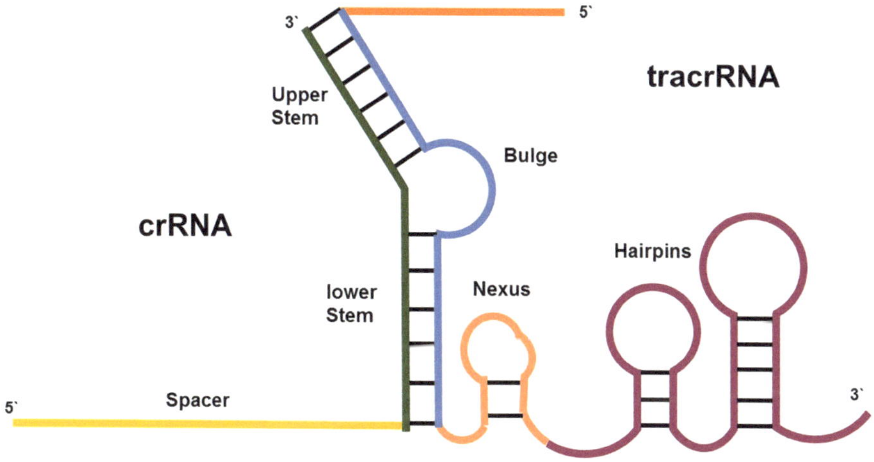

Figure 2.
Schematic structure of the crRNA:tracrRNA complex.

The major function of the sgRNA is efficiently the target region detection through the PAM sequence to edit a gene precisely. However, two vital challenges include efficacy and specificity for designing an effective sgRNA [28]. According to the significance of the sgRNA function, the role and designing tools of sgRNA will be outlined in the following.

2. The role of sgRNA in CRISPR technology

CRISPR/Cas genome editing technology can cause double-stranded DNA breaks (DSBs) in predefined genomic loci [29]. These DSBs are then repaired by the DNA repair systems of the target organism, which inherently can cause a mutation in the target gene. Despite the general processes driving all these genome editing systems being similar, CRISPR/Cas technology has emerged as the preferred technique due to its easy usage, low cost, outstanding adaptability, and ability to target several genes at once [30–32].

This technology consists of a Cas endonuclease, responsible for eliciting the DSB, and a short noncoding about sgRNA (20-nt), directing Cas to the correct genomic region for targeted genome editing. A chimeric gRNA (complementary to the target area) and trans-activating CRISPR-RNA are usually included in this sgRNA. Most Cas systems require the predesigned sgRNA to anneal immediately upstream of a PAM, which in the case of SpCas9 (the most extensively employed Cas protein for genome editing) is 5` NGG3` [4, 5, 33]. In these cases, the PAM is required to cleave target DNA around 3-nt upstream of this region. DSBs are repaired by either non-homologous end joining (NHEJ) or homology-directed repair (HDR) as two central intrinsic DNA repair systems [34]. The error-prone nature of NHEJ is the main DNA repair route in species and the most common and straightforward pathway in genome

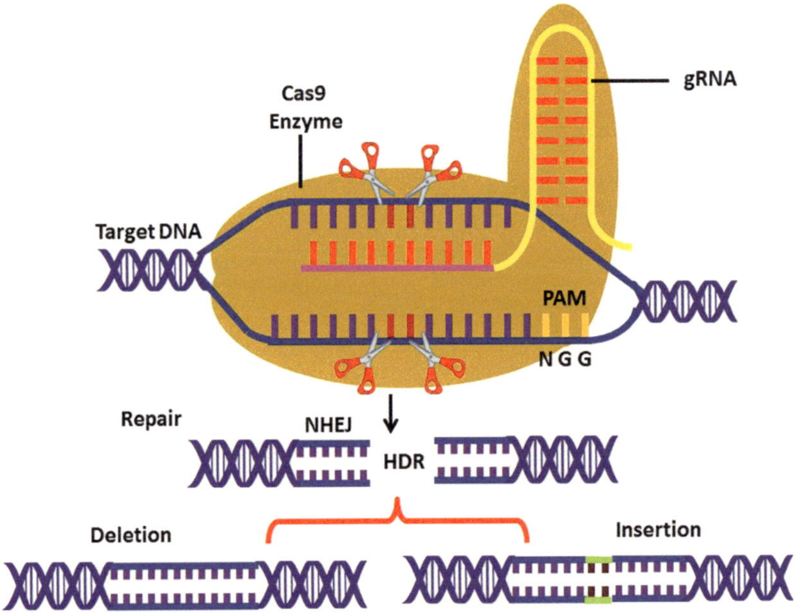

Figure 3.
CRISPR/Cas9 technology for gene editing. The Cas9 DNA endonuclease is recruited by a single-guide RNA (sgRNA) that detects a genomic sequence followed by a 5′-NGG-3′ PAM motif.

System	ZFNs	TALENs	CRISPR
Function	Cleavage mediated by DNA-protein interaction	Cleavage mediated by DNA-protein interaction	Cleavage mediated by DNA-RNA interaction
Nuclease designing and assembly	Hard and costly	Mostly possible in the laboratory, but highly difficult	Easy
Designing efficiency	Low	High	High
Assembly efficiency	Variable	High (%99<)	High (%90<)
Targeting range	Restricted, because of dependence on ZF modules	Unrestricted because of independence on PAM	Restricted by PAM, but generally unrestricted
Off-target effects	Yes	Yes	Yes
Sensitivity to DNA methylation	Undefined	Sensitive to CpG methylation	Nonsensitive to CpG methylation
High operating power	No	Restricted	Possible

Table 1.
Comparison between genome editing systems.

editing, causing small insertions or deletions (indels) to interrupt the target sequence (**Figure 3**) [35].

CRISPR/Cas technology relies on DNA-RNA interaction as well as simple design of RNA molecule for each specific sequence. However, protein-DNA interaction also depends on zinc-finger nucleases (ZFNs) and transcription activator-like effector nucleases (TALENs). So there is a required reconstruction for each target DNA sequence. This is a significant benefit of CRISPR/Cas technology [36]. In addition, the technology has three other advantages over TALENs and ZFNs, including the following (**Table 1**):

1. *Simplicity*: The 20 nt sequence of sgRNA can be simply designed to target any sequence [5].

2. *Multifunction*: As this technology's main advantage over TALENs and ZFNs, several sgRNAs can function differently and simultaneously on variable genomic sites [37].

3. *Non-specificity to the target DNA methylation*: the CRISPR/Cas technology can edit some genome sites which are highly regulated by epigenetic changes [38], especially in plants that contain around 70% of CpG/CpNpG sites and CpG islands methylated in the proximal exons promoter [39].

2.1 The role of PAM

Although specific targeting significantly depends on the sgRNA sequence, PAM-specific sequence plays a crucial role in the effective enzymatic activity

Bacterial species	Endonuclease	PAM-specific sequence
Streptococcus pyogenes	SpCas9	NGG
	SpCas9 D1135E	NAG
	SpCas9 VRER	NGCG
	SpCas9 EQR	NGAG
	SpCas9 VQR	NGNG/ NGAN
Staphylococcus aureus	SaCas9	NNGRR(N)/ NNGRRT
Neisseria meningitidis	NmCas9	NNNNGATT
Streptococcus thermophilus	StCas9	NNAGAAW
Treponema denticola	TdCas9	NAAAAC
S. aureus	SaCas12a	TTN

N = A,T,G,C; R = G,A; and W = A, T.

Table 2.
Different PAM sequences of some Cas endonucleases and their origins.

of Cas endonuclease. In this system, Cas9 endonuclease can cleave any genome sequence immediately located on five nucleotides (nt) downstream of PAM sequence [25]. PAM sequences recognized by SpCas9 and StCas9 (from *Streptococcus thermophilus*) is, respectively, 3`-NGG-5` and 5`-NNAGAAW-3` (**Table 2**) [40, 41]. The SpCas9-sgRNA complex first seeks the complement sequence of PAM in the target genome and then the sgRNA base pairs to the target DNA, then the DNA is cleaved by SpCas9 to create DSB. Generally, the length of the DNA detection sequence in the crRNA region is 20 nt, though the more base pairs bind between RNA and DNA, the more the specificity of the sgRNA function can enhance. Hence, the 20 nt sequence of sgRNA and 3 nt in PAM play key roles in the specific targeting of CRISPR/Cas technology. However, there are some limitations in using the 3`-NGG-5` motif, particularly at high AT sequences of the target genome [42, 43].

2.2 Variety of sgRNA types

According to the development of the CRISPR system as technology and sgRNA invention, there are some improvements of sgRNA to enhance the efficiency, precision, and specificity of the genome editing technology [44]. The improvements include as follows:

1. *Truncated guide RNA (tuRNA)*: the RNA contains a homologous 17 nt sequence to the target gene. Hence, the specificity of Cas endonuclease activity can increase by reducing off-target sequences [45].

2. *Polycistronic tRNA-gRNA (PTG/Cas9)*: in the system, the RNA is a frequent repetition of the tRNA-gRNA units and target-specific spacer sequences for targeting several sites in the genome sequence [46]. After PTG transcription, the primary copy is matured as sgRNAs through tRNA processing system and RNaseP and RNaseZ activity. In addition, the sgRNAs can interact with Cas endonuclease to target multiple genes specifically and simultaneously [47].

2.3 Off-target effect

Off-target mutation is a major challenge in CRISPR/Cas technology [48]. If the gRNA sequence contains less than three heterologous nucleotides to an off-target region, off-target effects will be observed [49]. Studies have indicated that mismatched pairs at the end of the 3` terminal of the target sequence are not tolerated (typically 8–14 nucleotides upstream of PAM sequence). In contrast, the mismatched pairs at the 5` terminal of the target sequence are more tolerable [50]. The sgRNA/Cas9 also can influence the off-target effects [49].

Generally, although Cas9 protein can be differently used according to high endonuclease activity and the wide targeting range of the enzyme, the high molecular weight of the Cas9 endonuclease and off-target effects can restrict the popularity of the enzyme. Nevertheless, some variants of Cas9, such as SpCas9-HF and eSpCas9 [51, 52], can be mutated. So, the mutation can reduce the nonspecific interaction between the Cas9 protein and the target sequence. Digenome-seq, GUIDE-Seq (Genome-wide, Unbiased Identification of DSBs Enabled by Sequencing), and HTGTS (High-Throughput Genome-Wide Translocation Sequencing) can be employed for detecting off-target regions [53]. However, precisely designing sgRNA can significantly decrease the rate of the off-target effects [44].

3. Characteristics of an effective sgRNA

In addition to directing Cas endonuclease, sgRNA can stimulate Cas endonuclease activity. These functions of sgRNA can clarify how to tackle on-target effects [54]. It has been demonstrated that the proximal and distal ends of the PAM sequence are highly responsible for improving on-target effects. Besides, genomic frameworks of the target sequence, GC content, sgRNA length, and secondary structure play significant roles in enhancing the on-target rate [44]. Also, 5` terminal 20 nt of the sgRNA is highly efficient for on-target efficiency. However, there is insufficient information on the correlation between structure and sequence properties of sgRNA influencing the on-target effect [55]. At least, enhancement of the on-target rate can improve the efficiency of the CRISPR technology and facilitate the statistical interpretation of the edited genes rate [56].

3.1 GC content of the sgRNA

As mentioned above, the GC content of the sgRNA is closely related to the on-target rate and the efficiency of the CRISPR gene editing. It has been indicated that too high or low GC content is unsuitable for achieving high rate of the on-target effects [57]. The knockout effectiveness of the CRISPR/Cas9 system was significantly improved by changing the sgRNA structure by expanding the duplex length (about 5 nt) and replacing the fourth T by C or G [58]. Many studies have reported that GC content plays a key role in improving the knockout efficiency of the CRISPR/Cas9 system. The effective rate of GC content is 40–60 percent. It is also recommended that sgRNA containing 50 percent GC content is efficient for CRISPR gene editing [59–61]. However, some studies suggest higher than 60 percent GC content for each organism, such as *Escherichia coli* (62.5%) and *Vitis vinifera* (65%) [62, 63]. In addition to GC content, purine residues and curvature in positions C3 and C16 of sgRNA could be effective in improving on-target activity [64]. It has also been reported that sgRNAs containing 4 GC in the 6 nt close to the PAM sequence can effectively reduce off-target effects [65].

3.2 The sgRNA length

The most common sgRNA length is about 100 nt. Therefore, 5` terminal 20 nt of the sgRNA can be designed as the complement sequence of the target gene to direct precisely Cas endonuclease for achieving effective gene editing [66]. According to several studies, the less the sgRNA length is, the higher the rate of off-target effects increase [67–69]. However, as mentioned above, 17 nt length for tuRNA can be highly effective in reducing off-target effects. However, the length of the sgRNA recognition site is less than 15 nt, and Cas endonuclease will not show any activity [70, 71]. In addition to the sgRNA length, the efficiency and specificity of Cas cleavage activity in the targeting sequence are significantly influenced by the distance between the PAM site and the start codon [72].

3.3 The sgRNA secondary structure

The sgRNA secondary structure is highly responsible for effective Cas-target sequence binding [73]. There are also many reports to indicate that the presence of the quad stem-loop structure of sgRNA is a key factor in improving the efficiency of the riboprotein function. The repeat and anti-repeat region (stem-loop RAR, GAAA) can activate sgRNA processing before Cas-sgRNA binding. Besides, loops 2 (GAAA) and 3 (AGU) are demanded to create a stable riboprotein complex, but there is no report on the possible loop 1 role in sgRNA efficiency [74–77]. Besides, the hairpin structure of the sgRNA, particularly the inner side of the hairpin, plays a key role in cleaving target DNA by Cas9/sgRNA. In fact, the hairpin structure can provide a suitable conformation to bind Cas9 enzyme. If the loop-stem structure is elongated, the gene editing efficiency will be enhanced [78, 79]. It has been indicated that CRISPR efficiency can be improved when an engineered hairpin structure is inserted into the spacer region of sgRNA. The modified sgRNA can positively influence Cas-mediated transactivation and improve the function of the five different Cas9 and Cas12a variants. The evidence can demonstrate the effect of sgRNA secondary structure on the success rate of gene editing [80, 81]. There is also a correlation between the sgRNA secondary structure and the efficiency of the Cas9-mediated CRISPR [55]. It has been reported that sgRNA refolding can refine the destructive bonds of the deactivated sgRNAs. Also, heating or slowly chilling can thermodynamically activate these sgRNAs to improve Cas9 cleavage activity. At least, the sgRNA secondary structure can change the guide sequence activity and deactivated sgRNA can be recovered by refolding [82, 83].

3.4 The sgRNA sequence

In addition to these criteria, the sgRNA sequence features can show different efficiency levels. It has been reported that the functional sgRNA can be significantly accessed at certain nucleotide positions more than nonfunctional sgRNA. Particularly, 3` terminal nucleotides (positions 18–20) of the sgRNA can highly make a prominent difference in accessibility [84, 85]. The sgRNA 3` terminus, known as the seed region, is a key player in recognizing the target sequence. Therefore, accessibility of the final three bases of the seed region is a remarkable characteristic in distinguishing functional sgRNAs from nonfunctional ones based on structural analysis [86, 87]. Also, G at the 5` terminus of the spacer is demanded in the non-ribosomal and ribosomal complexes. Besides, G is extremely proper at positions −1 and − 2, close to PAM associated with the sequence preference for loading Cas9 endonuclease [88]. According to the finding that multiple U in the spacer results in low sgRNA expression, T is not preferred at the four positions nearest to the PAM. The downstream nucleotides of the PAM cooperate with the efficiency of sgRNA, while upstream sequences of the spacer do not show significant effects. The early termination of sgRNA transcription is mostly responsible for the reduced sgRNA expression rates caused by the high frequency of nonconsecutive T clusters in the protospacer. Cytosine is favored at the −3 position as the cleavage site of the sgRNA/Cas9 complex.

Additionally, guanines are favored from positions −14 to −17, while adenines are favored positions −5 to −12 [89–93]. Most molecular characteristics that govern sgRNA stability, loading, and targeting *in vivo* are yet unknown. While variable Cas9 off-target binding, positioning of the nucleosome, and sgRNA loading are not key factors, adenine depletion and guanine enrichment improved sgRNA activity and stability. There is also a close correlation between sgRNA efficiency and guanine enrichment PAM-proximal site, supposedly caused by G-quadruple structure increasing sgRNA stability [94–97].

4. Computational tools for designing sgRNA

Generally, the potential off-target effect is still a vital concern for several applications of CRISPR technology. There are many strategies including Cas endonucleases engineering, transcriptome analysis, tunable systems (small-molecule induction of Cas9, light-activated and intein-inactivated Cas9), functional screening after dCas9 treatment, direct delivery (RNP complex), truncated sgRNAs (small-guide RNAs), and separate Cas9-binding approaches (paired Cas9 nickase) to reduce the off-target activity in CRISPR/Cas gene editing (**Table 3**) [105–107]. However, designing sgRNA can be the most simple, effective, low-cost, and time-saving approach [28].

There are some key factors for designing an efficient sgRNA for CRISPR editing and reducing off-target effects. First, the GC contact should be 40–80 per cent, although a higher percentage is more desirable and beneficial [28]. Second, the sgRNA length needs to be 17–20 nt, depending on the used Cas enzyme. The shorter the sequence is designed, the less the off-target effects are observed; however, too short sequences can increase the off-target effects [71]. As the third factor, off-target effects may result from mismatches between sgRNA and the target sequence, according to the mismatch numbers and positions [96]. ΔG calculation provides a significant benefit to assess sgRNA-DNA binding potential. The ΔG of a highly effective gRNA is from −64.53 to −47.09 kcal/mol, but higher ΔG can increase mismatching rates, causing a high rate of the off-target effect [93]. Although the more the negative ΔG would be, the more stable the secondary structure of the sgRNA is observed, higher ΔG (~ −30 to −20 kcal/mol) can positively influence sgRNA transcription and practically make more effective RNA-RNA binding for a more functional secondary structure of the sgRNA [28].

According to all these criteria, designing sgRNA is a crucial concern in CRISPR technology [72]. As CRISPR/Cas contain two key players, including Cas endonuclease and sgRNA, to cooperate with genome editing, the development of each component could be beneficial to enhancing CRISPR/Cas editing. However, enzyme engineering

Strategies	Advantages	Limitation	Ref.
Cas engineering	PAMs improvement, shortening sgRNAs	Costly, time-consuming, reforming for each propose	[98]
Transcriptome analysis	Precisely detecting on- and off-target activity	Costly, time-consuming	[99]
Tunable system	Regulating Cas 9 working time, reducing undesirable DNA cleavage	Slow on-rate, decreased editing efficiency, time-consuming for inducing by light or chemicals	[100]
Functional screening	Validation of gene functions, controlling genetic disruption	Just applicable for the low-throughput formats	[101]
RNP complex	Reduced off-target effects, transient genome editing	High molecular weight, inducing phospholipid bilayer stress	[102]
Truncated sgRNA	Minimizing off-target effects without reducing on-target activity, decreasing sgRNA length	Editing in some new off-target sequences	[103]
Paired Cas9 nickase	High efficiency and specificity, using for insertion in target gene by NHEJ	Requiring two simultaneous sgRNAs	[104]

Table 3.
Benefits and limitations of the strategies predicting off-target effects.

is a costly, time-consuming, and complicated strategy. So, sgRNA designing could be more effective [108, 109]. Furthermore, an effective sgRNA should simultaneously show the highest on-target efficiency and the lowest off-target activity. So, several well-developed computational tools can be found to design sgRNA for high-efficiency genome editing [110]. In addition to the simplicity, high efficiency, and cost-effectiveness, the *in silico* tool offers adaptability, automation, and fast processing to analyze many genes [111].

In the last decade, many *in silico* tools have been introduced to developing CRISPR technology because there has been an urgent demand to design an effective sgRNA to create precise mutations via CRISPR/Cas. Some tools have combined several scoring methods and/or algorithms to provide better design services [112, 113]. In addition to the different features of effectively designing sgRNA, these tools would be user-friendly [114]. The most popular computational sgRNA designing tools are outlined in the following and the other in silico tools are summarized in **Table 4**. The outlined tools are able to offer candidate sgRNAs and simultaneously score off-target activity. Besides, they are more user-friendly and fast-processing tools [28, 44].

4.1 CHOPCHOP

CHOPCHOP website, one of the most conventional *in silico* tools for detecting target sequences, includes a clear interface and complete functions. There are more than 200 reference genomes on the website so that the users can search for the target sequence, genomic coordinates, and name of the desired gene. The users are also able to choose two different methods to detect off-target and seven scoring approaches for on-target efficiency before sgRNA designing (**Table 5**) [117, 125, 126].

Name	Services	URL	Ref.
sgRNA Designer	sgRNA designing	https://portals.broadinstitute.org/gpp/public/analysis-tools/sgrna-design	[115]
CRISPR-ERA	sgRNA designing	http://CRISPR-ERA.stanford.edu	[116]
CHOPCHOP	Detecting target sequence	https://chopchop.cbu.uib.no/about	[117]
CRISPRscan	sgRNA designing and analysis	https://www.crisprscan.org/	[118]
CRISPR-GE	sgRNA designing detecting target sequence	http://skl.scau.edu.cn/	[119]
CRISPR RGEN Tools	Detecting off-target sites	http://www.rgenome.net/	[120]
sgRNAcas9	sgRNA designing predicting off-target sites	http://biootools.com/	[121]
CRISPR MultiTargeter	Multiple sgRNA designing tool	http://multicrispr.net/	[122]
CRISPR-P	sgRNA designing for plants	http://crispr.hzau.edu.cn/CRISPR2/	[123]
CRISPOR	sgRNA designing Detecting on- and off-target sites	http://crispor.tefor.net/	[124]

Table 4.
Common ***in silico*** *tools for sgRNA designing in CRISPR technology.*

4.2 CRISPR RGEN tools

CRISPR/Cas-derived RNA-guided engineered nuclease (CRISPR RGEN) can provide several computational tools and sgRNA/Cas libraries, including nine tools such as Cas-OFFinder, Cas-Designer, and Digenome-Seq. Compared with other tools, Cas-Designer can rapidly detect potential off-target sites containing a DNA or RNA bulge. Besides, Cas-Designer can offer an out-of-frame score for each sgRNA to find the proper sites for the gene knockout [120, 127, 128]. Cas-OFFinder is also used to seek potential off-target positions of NmCas9 endonuclease (from *N.meningitides*), recognizing 5′-NNNNGMTT-3′ PAM sequence (M = A or C) as well as a 24-bp target sequence specific to the design sgRNA in the target genome. Also, mixed bases can be used by Cas-OFFinder to analyze the degeneracy of PAM sequences. At least, Cas-OFFinder can provide quick scanning for potential off-target positions in any sequenced genome, regardless of the mismatched nucleotides numbers or the PAM sequence limitation (**Table 5**) [120, 129, 130].

4.3 CRISPOR

Among these *in silico* tools for effectively designing sgRNA, CRISPOR includes various useful tools to design sgRNA, 417 genomes, and 19 PAM types. This *in silico* tool can receive genome coordinates and sequences with more than 2000 bp length as the inputs. After processing, comprehensive information is provided as the output. The result can be, by default, presented in two sections; first,

Tool	Covered genome	Nucleases	Nickase	Off-target analysis	Cas type
ChopChop	>200	Yes	No	Yes	Different Type II
CRISPR RGEN	350	Yes	No	Yes	Different Type II
CRISPOR	417	Yes	No	Yes	Different Type II

Table 5.
Comparing three ***in silico*** *tools.*

visualizing the PAM sites along the target sequence, available in different formats such as fasta, GenBank and SnapGene; second, providing a table containing all information such as 2 specificity scores and 10 efficiency scores for every predicted sgRNA (**Table 5**) [105, 124, 131, 132].

4.4 Challenges and limitations of *in silico* tools

Although the computational tools for sgRNA designing can facilitate on-target prediction and reduce off-target activity, some tools cannot cover all vital criteria. So, it is highly recommended to use different *in silico* tools. For example, some free, user-friendly, and reliable online tools include RNAfold and Mfold to predict sgRNA secondary structure [59, 79].

Moreover, all current prediction models struggle with four main challenges:

1. *Data insufficiency*: machine learning models (MLDs) operate better than other approaches due to their data-driven process. Nevertheless, they cannot accurately anticipate previously unseen data without sufficient data to fully extract features [133, 134].

2. *Unclear mechanism*: The mechanism of the CRISPR/Cas9 editing system has not been completely discovered and limits the characteristics used in the most recent cutting-edge algorithms. So, MDL-mediated approaches are not simply able to achieve significant advancements with sufficient data [133, 134].

3. *Data heterogeneity*: generated datasets from various platforms and cell types should be merged for data augmentation [133].

4. *Data imbalance*: the detection of the frequency of off-target sequences through high-throughput whole-genome sequencing is considerably less than the numbers detected by *in silico* prediction tools [133, 135].

5. Adenine base editors (ABEs)

Base editors (BEs), as chimeric proteins, contain a catalytic domain and a DNA targeting modules which is able to deaminate adenines and cytosines. There is no need to make DSBs in DNA bases editing when BEs are used for base editing. So, these proteins can reduce the off-target effects and random indels at the on-target sequences. The BEs have been introduced as novel promising tools to make precise gene modification [6, 136, 137]. ABEs are the fused Cas9 nickase with a deaminase domain converting A-G and C-T (C > T) at the target sequence [138]. In fact, ABEs can effectively and precisely convert A-G and C-T base pairs at the target site within the editing frame while producing few by-products, consequently reducing off-target activity significantly. The ABE variants can improve the precision of adenine base editing by reducing the off-target activities of RNA and DNA [105, 139]. It has been reported that ABEs cooperate to induce free-sgRNA transcriptome editing. ABEs have also been discovered to display RNA off-target activity and the capacity to edit their own transcripts [140, 141]. ABEs can generally produce significant off-target single-nucleotide variations (SNVs) in RNA sequences. Therefore, deaminase engineering enables ABE variants to decrease off-target mutation of SNVs in RNA sequences while increasing on-target efficiency with DNA [105, 142].

6. Conclusion

Over the last 30 years, genome editing technology, particularly CRISPR/Cas, has promoted biosciences by editing and targeting the genomic DNAs of any species. CRISPR/Cas is the most precise, effective, and affordable among all these genome editing technologies. Although there are diverse types and classes of CRISPR/Cas systems, they are not all applicable due to the high rate of off-target effects. Different approaches have been developed to decrease the off-target effects for enhancing the precision and efficiency of the different CRISPR/Cas techniques. According to a refined reference genome, a well-designed sgRNA can support high on-target efficiency to create a precise and desirable mutation. Finally, it is highly recommended to consider the criteria as mentioned above, including GC content, length, secondary structure, and sequence, for designing an effective sgRNA to achieve high-precision CRISPR genome editing.

Acknowledgements

The chapter was funded by Amin Techno Gene Private Virtual Lab (NGO), Tehran, Iran.

Conflict of interest

The authors declare no conflict of interest.

Author details

Reza Mohammadhassan[1*], Sara Tutunchi[2], Negar Nasehi[3], Fatemeh Goudarziasl[4] and Lena Mahya[5]

1 Plant Sciences Department, Amin Techno Gene Private Virtual Lab (NGO), Tehran, Iran

2 Department of Medical Genetics, Shahid Sadoughi University of Medical Sciences, Yazd, Iran

3 Faculty of Biological Sciences, Department of Nanobiotechnology, Tarbiat Modares University, Tehran, Iran

4 Faculty of Biological Science, Department of Biochemistry, University of Mazandaran, Babolsar, Iran

5 Medical Sciences Department, Amino Techno Gene Private Virtual Lab (NGO), Tehran, Iran

*Address all correspondence to: rezarmhreza22@gmail.com

References

[1] Dhawan M, Sharma M, Grewal RS. CRISPR Systems: RNA-Guided defence mechanisms in Bacteria and Archaea. International Journal of Current Microbiology and Applied Sciences. 2015;**4**(6):187-200

[2] Makarova KS, Wolf YI, Iranzo J, Shmakov SA, Alkhnbashi OS, Brouns SJ, et al. Evolutionary classification of CRISPR–Cas systems: A burst of class 2 and derived variants. Nature Reviews Microbiology. 2020;**18**(2):67-83

[3] Koonin EV, Makarova KS, Wolf YI. Evolutionary genomics of defense systems in archaea and bacteria. Annual Review of Microbiology. 2017;**71**:233-261

[4] Adli M. The CRISPR tool kit for genome editing and beyond. Nature Communications. 2018;**9**(1):1-13

[5] Pickar-Oliver A, Gersbach CA. The next generation of CRISPR–Cas technologies and applications. Nature Reviews Molecular Cell Biology. 2019;**20**(8):490-507

[6] Komor AC, Kim YB, Packer MS, Zuris JA, Liu DR. Programmable editing of a target base in genomic DNA without double-stranded DNA cleavage. Nature. 2016;**533**(7603):420-424

[7] Hille F, Richter H, Wong SP, Bratovič M, Ressel S, Charpentier E. The biology of CRISPR-Cas: Backward and forward. Cell. 2018;**172**(6):1239-1259

[8] Min YL, Bassel-Duby R, Olson EN. CRISPR correction of Duchenne muscular dystrophy. Annual Review of Medicine. 2019;**70**:239-255

[9] Xu CF, Chen GJ, Luo YL, Zhang Y, Zhao G, Lu ZD, et al. Rational designs of in vivo CRISPR-Cas delivery systems. Advanced Drug Delivery Reviews. 2021;**168**:3-29

[10] Mohammadhassan R, Akhavan S, Mahmoudi A, Khalkhali A, Barzin R. Antiviral activity of Echinacea (Echinacea Purpurea). International Journal of Advanced Biotechnology and Research. 2016;7(4):1493-1497

[11] Hynes AP, Rousseau GM, Agudelo D, Goulet A, Amigues B, Loehr J, et al. Widespread anti-CRISPR proteins in virulent bacteriophages inhibit a range of Cas9 proteins. Nature Communications. 2018;**9**(1):1-10

[12] Doudna JA, Charpentier E. The new frontier of genome engineering with CRISPR-Cas9. Science. 2014;**2014**:346

[13] Hess GT, Tycko J, Yao D, Bassik MC. Methods and applications of CRISPR-mediated base editing in eukaryotic genomes. Molecular Cell. 2017;**68**(1):26-43

[14] Ran FA, Cong L, Yan WX, Scott DA, Gootenberg JS, Kriz AJ, et al. In vivo genome editing using Staphylococcus aureus Cas9. Nature. 2015;**520**(7546):186-191

[15] Wright AV, Nuñez JK, Doudna JA. Biology and applications of CRISPR systems: Harnessing nature's toolbox for genome engineering. Cell. 2016;**164**(1-2):29-44

[16] Leenay RT. Deciphering, communicating, and engineering the CRISPR PAM. Journal of Molecular Biology. 2017;**429**(2):177-191

[17] Meeske AJ. RNA guide complementarity prevents self-targeting

in type VI CRISPR systems. Molecular Cell. 2018;**71**(5):791-801

[18] Liu Y, Pinto F, Wan X, Peng S, Li M, Xie Z, et al. Reprogrammed tracrRNAs enable repurposing RNAs as crRNAs and detecting RNAs. bioRxiv. 2018;**2021**:1-28

[19] Liu Y, Pinto F, Wan X, Yang Z, Peng S, Li M, et al. Reprogrammed tracrRNAs enable repurposing of RNAs as crRNAs and sequence-specific RNA biosensors. Nature Communications. 2022;**13**(1):1-12

[20] Charpentier E, Richter H, van der Oost J, White MF. Biogenesis pathways of RNA guides in archaeal and bacterial CRISPR-Cas adaptive immunity. FEMS Microbiology Reviews. 2015;**39**(3):428-441

[21] Jiao C, Sharma S, Dugar G, Peeck NL, Bischler T, Wimmer F, et al. Noncanonical crRNAs derived from host transcripts enable multiplexable RNA detection by Cas9. Science. 2021;**372**(6545):941-948

[22] Reis AC, Halper SM, Vezeau GE, Cetnar DP, Hossain A, Clauer PR, et al. Simultaneous repression of multiple bacterial genes using nonrepetitive extra-long sgRNA arrays. Nature Biotechnology. 2019;**37**(11):1294-1301

[23] Nelles DA, Fang MY, O'Connell MR, Xu JL, Markmiller SJ, Doudna JA, et al. Programmable RNA tracking in live cells with CRISPR/Cas9. Cell. 2016;**165**(3):488-496

[24] Hirano H, Gootenberg JS, Horii T, Abudayyeh OO, Kimura M, Hsu PD, et al. Structure and engineering of Francisella novicida Cas9. Cell. 2016;**164**(5):950-961

[25] Gasiunas G, Young JK, Karvelis T, Kazlauskas D, Urbaitis T, Jasnauskaite M, et al. A catalogue of biochemically diverse CRISPR-Cas9 orthologs. Nature Communications. 2020;**11**(1):1-10

[26] Chyou TY. Prediction and diversity of tracrRNAs from type II CRISPR-Cas systems. RNA Biology. 2019;**16**(3):423-434

[27] Hiranniramol K, Chen Y, Liu W, Wang X. Generalizable sgRNA design for improved CRISPR/Cas9 editing efficiency. Bioinformatics. 2020;**36**(9):2684-2689

[28] Baghini SS, Gardanova ZR, Zekiy AO, Shomali N, Tosan F, Jarahian M. Optimizing sgRNA to improve CRISPR/Cas9 knockout efficiency: Special focus on human and animal cell. Frontiers in Bioengineering and Biotechnology. 2021;**2021**:9

[29] Mohanta TK, Bashir T, Hashem A. Genome editing tools in plants. Genes. 2017;**8**(12):399

[30] Van de Wiel CCM, Schaart JG, Lots LAP, Smulders MJM. New traits in crops produced by genome editing techniques based on deletions. Plant Biotechnology Reports. 2017;**11**(1):1-8

[31] Mishra R, Zhao K. Genome editing technologies and their applications in crop improvement. Plant Biotechnology Reports. 2018;**12**(2):57-68

[32] Vats S, Kumawat S, Kumar V, Patil GB, Joshi T, Sonah H, et al. Genome editing in plants: Exploration of technological advancements and challenges. Cell. 2019;**8**:11

[33] Knott GJ, Doudna JA. CRISPR-Cas guides the future of genetic engineering. Science. 2018;**361**(6405):866-869

[34] Yang H, Wu JJ, Tang T, Liu KD, Dai C. CRISPR/Cas9-mediated genome editing efficiently creates specific mutations at multiple loci using one

sgRNA in Brassica napus. Scientific Reports. 2017;**7**(1):1-13

[35] Musunuru K. The hope and hype of CRISPR-Cas9 genome editing: A review. JAMA Cardiology. 2017;**2**(8):914-919

[36] Raper AT, Stephenson AA, Suo Z. Functional insights revealed by the kinetic mechanism of CRISPR/Cas9. Journal of the American Chemical Society. 2018;**140**(8):2971-2984

[37] Ma X, Zhang Q, Zhu Q, Liu W, Chen Y, Qiu R, et al. A robust CRISPR/Cas9 system for convenient, high-efficiency multiplex genome editing in monocot and dicot plants. Molecular Plant. 2015;**8**(8):1274-1284

[38] Nadakuduti SS, Enciso-Rodríguez F. Advances in genome editing with CRISPR systems and transformation technologies for plant DNA manipulation. Frontiers in Plant Science. 2021;**11**:637159

[39] Kang JG, Park JS, Ko JH, Kim YS. Regulation of gene expression by altered promoter methylation using a CRISPR/Cas9-mediated epigenetic editing system. Scientific Reports. 2019;**9**(1):1-12

[40] Hao M, Cui Y, Qu X. Analysis of CRISPR-Cas system in Streptococcus thermophilus and its application. Frontiers in Microbiology. 2018;**9**:257

[41] Miller SM, Wang T, Randolph PB, Arbab M, Shen MW, Huang TP, et al. Continuous evolution of SpCas9 variants compatible with non-G PAMs. Nature Biotechnology. 2020;**38**(4):471-481

[42] Burgess SM. Genome editing by targeted nucleases and the CRISPR/Cas revolution. The Liver: Biology and Pathobiology. 2020;**2020**:953-964

[43] Fallahi S, Mohammadhassan R. A review of pharmaceutical recombinant proteins and gene transformation approaches in transgenic poultry. Journal of Tropical Life Science. 2020;**10**(2):163-173

[44] Liu G. Computational approaches for effective CRISPR guide RNA design and evaluation. Computational and Structural Biotechnology Journal. 2020;**18**:35-44

[45] Rose JC, Popp NA, Richardson CD, Stephany JJ, Mathieu J, Wei CT, et al. Suppression of unwanted CRISPR-Cas9 editing by co-administration of catalytically inactivating truncated guide RNAs. Nature Communications. 2020;**11**(1):1-11

[46] Xie K, Yang Y. A multiplexed CRISPR/Cas9 editing system based on the endogenous tRNA processing. In: Plant Genome Editing with CRISPR Systems. New York: Humana; 2019. pp. 63-73

[47] Hui L, Zhao M, He J, Hu Y, Huo Y, Hao H, et al. A simple and reliable method for creating PCR-detectable mutants in Arabidopsis with the polycistronic tRNA–gRNA CRISPR/Cas9 system. Acta Physiologiae Plantarum. 2019;**41**(10):1-14

[48] Vakulskas CA, Behlke MA. Evaluation and reduction of CRISPR off-target cleavage events. Nucleic Acid Therapeutics. 2019;**29**(4):167-174

[49] Kempton HR, Qi LS. When genome editing goes off-target. Science. 2019;**364**(6437):234-236

[50] Graham N, Patil GB, Bubeck DM, Dobert RC, Glenn KC, Gutsche AT, et al. Plant genome editing and the relevance of off-target changes. Plant Physiology. 2020;**183**(4):1453-1471

[51] Chen JS, Dagdas YS, Kleinstiver BP, Welch MM, Sousa AA, Harrington LB,

et al. Enhanced proofreading governs CRISPR–Cas9 targeting accuracy. Nature. 2017;**550**(7676):407-410

[52] Fan R, Chai Z, Xing S, Chen K, Qiu F, Chai T, et al. Shortening the sgRNA-DNA interface enables SpCas9 and eSpCas9 (1.1) to nick the target DNA strand. Life Sciences. 2020;**63**(11):1619-1630

[53] Tsai SQ, Zheng Z, Nguyen NT, Liebers M, Topkar VV, Thapar V, et al. GUIDE-seq enables genome-wide profiling of off-target cleavage by CRISPR-Cas nucleases. Nature Biotechnology. 2015;**33**(2):187-197

[54] Hajiahmadi Z, Movahedi A, Wei H, Li D, Orooji Y, Ruan H, et al. Strategies to increase on-target and reduce off-target effects of the CRISPR/Cas9 system in plants. International Journal of Molecular Sciences. 2019;**20**:15

[55] Jensen KT, Fløe L, Petersen TS, Huang J, Xu F, Bolund L, et al. Chromatin accessibility and guide sequence secondary structure affect CRISPR-Cas9 gene editing efficiency. FEBS Letters. 2017;**591**(13):1892-1901

[56] Wang D, Zhang C, Wang B, Li B, Wang Q, Liu D, et al. Optimized CRISPR guide RNA design for two high-fidelity Cas9 variants by deep learning. Nature Communications. 2019;**10**(1):1-14

[57] Tatiossian KJ, Clark RD, Huang C, Thornton ME, Grubbs BH, Cannon PM. Rational selection of CRISPR-Cas9 guide RNAs for homology-directed genome editing. Molecular Therapy. 2021;**29**(3):1057-1069

[58] Dang Y, Jia G, Choi J, Ma H, Anaya E, Ye C, et al. Optimizing sgRNA structure to improve CRISPR-Cas9 knockout efficiency. Genome Biology. 2015;**16**(1):1-10

[59] Schindele P, Wolter F, Puchta H. CRISPR guide RNA design guidelines for efficient genome editing. In: RNA Tagging. New York: Humana; 2020. pp. 331-342

[60] Dhanjal JK, Dammalapati S, Pal S, Sundar D. Evaluation of off-targets predicted by sgRNA design tools. Genomics. 2020;**112**(5):3609-3614

[61] Chang AY. Single-guide RNAs: Rationale and design. In: CRISPR Genome Surgery in Stem Cells and Disease Tissues. England: Academic Press; 2022. pp. 47-55

[62] Ren F, Ren C, Zhang Z, Duan W, Lecourieux D, Li S, et al. Efficiency optimization of CRISPR/Cas9-mediated targeted mutagenesis in grape. Frontiers in Plant Science. 2019;**10**:612

[63] Xu Hua F, Wainberg M, Kundaje A, Fire AZ. High-throughput characterization of Cascade type IE CRISPR guide efficacy reveals unexpected PAM diversity and target sequence preferences. Genetics. 2017;**206**(4):1727-1738

[64] Bruegmann T, Deecke K, Fladung M. Evaluating the efficiency of gRNAs in CRISPR/Cas9 mediated genome editing in poplars. International Journal of Molecular Sciences. 2019;**20**:15

[65] Liu Y, Wei WP, Ye BC. High GC content Cas9-mediated genome-editing and biosynthetic gene cluster activation in Saccharopolyspora erythraea. ACS Synthetic Biology. 2018;**7**(5):1338-1348

[66] Zhao C, Wang Y, Nie X, Han X, Liu H, Li G, et al. Evaluation of the effects of sequence length and microsatellite instability on single-guide RNA activity and specificity. International Journal of Biological Sciences. 2019;**15**(12):2641

[67] Karmakar S, Behera D, Baig MJ, Molla KA. In vitro Cas9 cleavage assay to check guide RNA efficiency. In: CRISPR-Cas Methods. New York: Humana; 2021. pp. 23-39

[68] Li J, Hong S, Chen W, Zuo E, Yang H. Advances in detecting and reducing off-target effects generated by CRISPR-mediated genome editing. Journal of Genetics and Genomics. 2019;**46**(11):513-521

[69] Zhu LJ. Overview of guide RNA design tools for CRISPR-Cas9 genome editing technology. Frontiers in Biology. 2015;**10**(4):289-296

[70] Lv J, Wu S, Wei R, Li Y, Jin J, Mu Y, et al. The length of guide RNA and target DNA heteroduplex effects on CRISPR/Cas9 mediated genome editing efficiency in porcine cells. Journal of Veterinary Science. 2019;**20**:3

[71] Zhang JP, Li XL, Neises A, Chen W, Hu LP, Ji GZ, et al. Different effects of sgRNA length on CRISPR-mediated gene knockout efficiency. Scientific Reports. 2016;**6**(1):1-10

[72] Matson AW, Hosny N, Swanson ZA, Hering BJ, Burlak C. Optimizing sgRNA length to improve target specificity and efficiency for the GGTA1 gene using the CRISPR/Cas9 gene editing system. PLoS One. 2019;**14**:12

[73] Xu J, Lian W, Jia Y, Li L, Huang Z. Optimized guide RNA structure for genome editing via Cas9. Oncotarget. 2017;**8**:55

[74] Hassan MM, Chowdhury AK, Islam T. In silico analysis of gRNA secondary structure to predict its efficacy for plant genome editing. In: CRISPR-Cas Methods. New York: Humana; 2021. pp. 15-22

[75] Liang G, Zhang H, Lou D, Yu D. Selection of highly efficient sgRNAs for CRISPR/Cas9-based plant genome editing. Scientific Reports. 2016;**6**(1):1-8

[76] Uniyal AP, Mansotra K, Yadav SK, Kumar V. An overview of designing and selection of sgRNAs for precise genome editing by the CRISPR-Cas9 system in plants. 3 Biotech. 2019;**9**(6):1-9

[77] Liang Y, Eudes A, Yogiswara S, Jing B, Benites VT, Yamanaka R, et al. A screening method to identify efficient sgRNAs in Arabidopsis, used in conjunction with cell-specific lignin reduction. Biotechnology for Biofuels. 2019;**12**(1):1-15

[78] Jiang M, Ye Y, Li J. Core Hairpin Structure of SpCas9 sgRNA functions in a sequence-and spatial conformation–dependent manner. Translating Life Sciences Innovation. 2021;**26**(1):92-102

[79] Dong C, Gou Y, Lian J. SgRNA engineering for improved genome editing and expanded functional assays. Current Opinion in Biotechnology. 2022;**75**:102697

[80] Kocak DD, Josephs EA, Bhandarkar V, Adkar SS, Kwon JB, Gersbach CA. Increasing the specificity of CRISPR systems with engineered RNA secondary structures. Nature Biotechnology. 2019;**37**(6):657-666

[81] Hu Z, Wang Y, Liu Q, Qiu Y, Zhong Z, Li K, et al. Improving the precision of base editing by bubble hairpin single guide RNA. MBio. 2021;**12**:2

[82] Konstantakos V, Nentidis A, Krithara A, Paliouras G. CRISPR–Cas9 gRNA efficiency prediction: An overview of predictive tools and the role of deep learning. Nucleic Acids Research. 2022;**50**(7):3616-3637

[83] Thyme SB, Akhmetova L, Montague TG, Valen E, Schier AF. Internal guide RNA interactions interfere with Cas9-mediated cleavage. Nature Communications. 2016;**7**(1):1-7

[84] Muhammad Rafid AH, Toufikuzzaman M, Rahman MS, Rahman MS. CRISPRpred (SEQ): A sequence-based method for sgRNA on target activity prediction using traditional machine learning. BMC Bioinformatics. 2020;**21**(1):1-13

[85] Wong N, Liu W, Wang X. WU-CRISPR: Characteristics of functional guide RNAs for the CRISPR/Cas9 system. Genome Biology. 2015;**16**(1):1-8

[86] Wilson LO, O'Brien AR, Bauer DC. The current state and future of CRISPR-Cas9 gRNA design tools. Frontiers in Pharmacology. 2018;**9**:749

[87] Gerashchenkov GA, Rozhnova NA, Kuluev BR, Kiryanova OY, Gumerova GR, Knyazev AV, et al. Design of guide RNA for CRISPR/Cas plant genome editing. Molecular Biology. 2020;**54**(1):24-42

[88] Horodecka K, Düchler M. CRISPR/Cas9: Principle, applications, and delivery through extracellular vesicles. International Journal of Molecular Sciences. 2021;**22**:11

[89] Hanna RE. Design and analysis of CRISPR–Cas experiments. Nature Biotechnology. 2020;**38**(7):813-823

[90] Kim N, Kim HK, Lee S, Seo JH, Choi JW, Park J, et al. Prediction of the sequence-specific cleavage activity of Cas9 variants. Nature Biotechnology. 2020;**38**(11):1328-1336

[91] Moreb EA, Lynch MD. Genome dependent Cas9/gRNA search time underlies sequence dependent gRNA activity. Nature Communications. 2021;**12**(1):1-13

[92] Xiang X, Corsi GI, Anthon C, Qu K, Pan X, Liang X, et al. Enhancing CRISPR-Cas9 gRNA efficiency prediction by data integration and deep learning. Nature Communications. 2021;**12**(1):1-9

[93] Corsi GI, Qu K, Alkan F, Pan X, Luo Y, Gorodkin J. CRISPR/Cas9 gRNA activity depends on free energy changes and on the target PAM context. Nature Communications. 2022;**13**(1):1-14

[94] Sledzinski P, Nowaczyk M, Olejniczak M. Computational tools and resources supporting CRISPR-Cas experiments. Cell. 2020;**9**:5

[95] Li C, Chu W, Gill RA, Sang S, Shi Y, Hu X, et al. Computational tools and resources for CRISPR/Cas genome editing. Genomics, Proteomics and Bioinformatics. 2022 (In Press). DOI: 10.1016/j.gpb.2022.02.006

[96] Luo M, Wang J, Dong Z, Wang C, Lu G. CRISPR-Cas9 sgRNA design and outcome assessment: Bioinformatics tools and aquaculture applications. Aquaculture and Fisheries. 2021;**7**(2):121-130

[97] Aslam MA, Hammad M, Ahmad A, Altenbuchner J, Ali H. Delivery methods, resources and design tools in CRISPR/Cas. In: CRISPR Crops. Singapore: Springer; 2021. pp. 63-116

[98] Slaymaker IM, Gaudelli NM. Engineering Cas9 for human genome editing. Current Opinion in Structural Biology. 2021;**69**:86-98

[99] Hrdlickova R, Toloue M, Tian B. RNA-Seq methods for transcriptome

analysis. Wiley Interdisciplinary Reviews: RNA. 2017;**8**:1

[100] Nuñez JK, Harrington LB, Doudna JA. Chemical and biophysical modulation of Cas9 for tunable genome engineering. ACS Chemical Biology. 2016;**11**(3):681-688

[101] Sun N, Petiwala S, Wang R, Lu C, Hu M, Ghosh S, et al. Development of drug-inducible CRISPR-Cas9 systems for large-scale functional screening. BMC Genomics. 2019;**20**(1):1-15

[102] Zhang S, Shen J, Li D, Cheng Y. Strategies in the delivery of Cas9 ribonucleoprotein for CRISPR/Cas9 genome editing. Theranostics. 2021;**11**(2):614

[103] Cui Z, Tian R, Huang Z, Jin Z, Li L, Liu J, et al. FrCas9 is a CRISPR/Cas9 system with high editing efficiency and fidelity. Nature Communications. 2022;**13**(1):1-12

[104] Gopalappa R, Suresh B, Ramakrishna S, Kim H. Paired D10A Cas9 nickases are sometimes more efficient than individual nucleases for gene disruption. Nucleic Acids Research. 2018;**46**(12):e71

[105] Manghwar H, Li B, Ding X, Hussain A, Lindsey K, Zhang X, et al. CRISPR/Cas systems in genome editing: Methodologies and tools for sgRNA design, off-target evaluation, and strategies to mitigate off-target effects. Advanced Science. 2020;7(6):1902312

[106] Asmamaw M, Zawdie B. Mechanism and applications of CRISPR/Cas-9-mediated genome editing. Biologics: Targets and Therapy. 2021;**15**:353

[107] Nasir MA, Nawaz S, Huang J. A review: Computational approaches to design sgRNA of CRISPR-Cas9. Current Bioinformatics. 2022;**17**(1):2-18

[108] Xue L, Tang B, Chen W, Luo J. Prediction of CRISPR sgRNA activity using a deep convolutional neural network. Journal of Chemical Information and Modeling. 2018;**59**(1):615-624

[109] Xiao LM, Wan YQ, Jiang ZR. AttCRISPR: A spacetime interpretable model for prediction of sgRNA on-target activity. BMC Bioinformatics. 2020;**22**(1):1-17

[110] Chuai GH, Wang QL, Liu Q. In silico meets in vivo: Towards computational CRISPR-based sgRNA design. Trends in Biotechnology. 2017;**35**(1):12-21

[111] Karlapudi AP, Venkateswarulu TC, Tammineedi J, Srirama K, Kanumuri L, Kodali VP. In silico sgRNA tool design for CRISPR control of quorum sensing in Acinetobacter species. Genes and Diseases. 2018;**5**(2):123-129

[112] Choudhary S, Ubale A, Padiya J, Mikkilineni V. Application of bioinformatics tools in CRISPR/Cas. In: CRISPR/Cas Genome Editing. Cham: Springer; 2020. pp. 31-52

[113] Peng H, Zheng Y, Zhao Z, Li J. Multigene editing: Current approaches and beyond. Briefings in Bioinformatics. 2021;**22**:5

[114] Zhou H, Zhou M, Li D, Manthey J, Lioutikova E, Wang H, et al. Whole genome analysis of CRISPR Cas9 sgRNA off-target homologies via an efficient computational algorithm. BMC Genomics. 2017;**18**(9):31-38

[115] Doench JG, Fusi N, Sullender M, Hegde M, Vaimberg EW, Donovan KF, et al. Optimized sgRNA design to maximize activity and minimize off-target effects

of CRISPR-Cas9. Nature Biotechnology. 2016;**34**(2):184-191

[116] Liu H, Wei Z, Dominguez A, Li Y, Wang X, Qi LS. CRISPR-ERA: A comprehensive design tool for CRISPR-mediated gene editing, repression and activation. Bioinformatics. 2015;**31**(22):3676-3678

[117] Montague TG, Cruz JM, Gagnon JA, Church GM, Valen E. CHOPCHOP: A CRISPR/Cas9 and TALEN web tool for genome editing. Nucleic Acids Research. 2014;**42**(W1):W401-W407

[118] Moreno-Mateos MA, Vejnar CE, Beaudoin JD, Fernandez JP, Mis EK, Khokha MK, et al. CRISPRscan: Designing highly efficient sgRNAs for CRISPR-Cas9 targeting in vivo. Nature Methods. 2015;**12**(10):982-988

[119] Xie X, Ma X, Zhu Q, Zeng D, Li G, Liu YG. CRISPR-GE: A convenient software toolkit for CRISPR-based genome editing. Molecular Plant. 2017;**10**(9):1246-1249

[120] Hwang G-H, Kim J-S, Bae S. Web-based CRISPR toolkits: Cas-OFFinder, cas-designer, and cas-analyzer. In: CRISPR Guide RNA Design. New York, NY: Springer; 2021. pp. 23-33

[121] Xie S, Shen B, Zhang C, Huang X, Zhang Y. sgRNAcas9: A software package for designing CRISPR sgRNA and evaluating potential off-target cleavage sites. PloS One. 2014;**9**(6):e100448

[122] Prykhozhij SV, Rajan V, Gaston D, Berman JN. CRISPR multitargeter: A web tool to find common and unique CRISPR single guide RNA targets in a set of similar sequences. PLoS One. 2015;**10**(3):e0119372

[123] Lei Y, Lu L, Liu HY, Li S, Xing F, Chen LL. CRISPR-P: A web tool for synthetic single-guide RNA design of CRISPR-system in plants. Molecular Plant. 2014;7(9):1494-1496

[124] Concordet J-P, Haeussler M. CRISPOR: Intuitive guide selection for CRISPR/Cas9 genome editing experiments and screens. Nucleic Acids Research. 2018;**46**(W1):W242-W245

[125] Labun K, Montague TG, Gagnon JA, Thyme SB, Valen E. CHOPCHOP v2: A web tool for the next generation of CRISPR genome engineering. Nucleic Acids Research. 2016;**44**(W1):W272-W276

[126] Labun K, Krause M, Torres Cleuren Y, Valen E. CRISPR genome editing made easy through the CHOPCHOP website. Current Protocols. 2021;**1**(4):e46

[127] Park J, Bae S, Kim J-S. Cas-Designer: A web-based tool for choice of CRISPR-Cas9 target sites. Bioinformatics. 2015;**31**(24):4014-4016

[128] Bradford J, Perrin D. A benchmark of computational CRISPR-Cas9 guide design methods. PLoS Computational Biology. 2019;**15**(8):e1007274

[129] Bae S, Park J, Kim J-S. Cas-OFFinder: A fast and versatile algorithm that searches for potential off-target sites of Cas9 RNA-guided endonucleases. Bioinformatics. 2014;**30**(10):1473-1475

[130] Zhao C et al. CRISPR-offinder: A CRISPR guide RNA design and off-target searching tool for user-defined protospacer adjacent motif. International Journal of Biological Sciences. 2017;**13**(12):1470

[131] Haeussler M et al. Evaluation of off-target and on-target scoring algorithms and integration into the guide RNA selection tool CRISPOR. Genome Biology. 2016;**17**(1):1-12

[132] Haeussler M. CRISPR off-targets: A question of context. Cell Biology and Toxicology. 2020;**36**(1):5-9

[133] Wang J, Zhang X, Cheng L, Luo Y. An overview and metanalysis of machine and deep learning-based CRISPR gRNA design tools. RNA Biology. 2020;**17**(1):13-22

[134] Jin J, Huo L, Song S, Ajani JA. CRISPR/Cas9 Technology Improvements and the RNA-Editing Trend. Journal of Molecular Biology and Biotechnology. 2020;**5**:5

[135] Li W, Wang XB, Xu Y. Recognition of CRISPR Off-target cleavage sites with SeqGAN. Current Bioinformatics. 2022;**17**(1):101-107

[136] Kim D, Kim DE, Lee G, Cho SI, Kim JS. Genome-wide target specificity of CRISPR RNA-guided adenine base editors. Nature Biotechnology. 2019;**37**(4):430-435

[137] Jin S, Zong Y, Gao Q, Zhu Z, Wang Y, Qin P, et al. Cytosine, but not adenine, base editors induce genome-wide off-target mutations in rice. Science. 2019;**364**(6437):292-295

[138] Gaudelli NM, Komor AC, Rees HA, Packer MS, Badran AH, Bryson DI, et al. Programmable base editing of A• T to G• C in genomic DNA without DNA cleavage. Nature Biotechnology. 2017;**551**(7681):464-471

[139] Rees HA, Wilson C, Doman JL, Liu DR. Analysis and minimization of cellular RNA editing by DNA adenine base editors. Science Advances. 2019;**5**:5

[140] Grünewald J, Zhou R, Garcia SP, Iyer S, Lareau CA, Aryee MJ, et al. Transcriptome-wide off-target RNA editing induced by CRISPR-guided DNA base editors. Nature. 2019;**569**(7756):433-437

[141] Grünewald J, Zhou R, Iyer S, Lareau CA, Garcia SP, Aryee MJ, et al. CRISPR DNA base editors with reduced RNA off-target and self-editing activities. Nature Biotechnology. 2019;**37**(9):1041-1048

[142] Zhou C, Sun Y, Yan R, Liu Y, Zuo E, Gu C, et al. Off-target RNA mutation induced by DNA base editing and its elimination by mutagenesis. Nature. 2019;**571**(7764):275-278

Chapter 5

Recent Advances in *In Vivo* Genome Editing Targeting Mammalian Preimplantation Embryos

Masahiro Sato, Masato Ohtsuka, Emi Inada, Shingo Nakamura, Issei Saitoh and Shuji Takabayashi

Abstract

CRISPR-based genome engineering has been widely used for producing gene-modified animals such as mice and rats, to explore the function of a gene of interest and to create disease models. However, it always requires the *ex vivo* handling of preimplantation embryos, as exemplified by the microinjection of genome editing components into zygotes or *in vitro* electroporation of zygotes in the presence of genome editing components, and subsequent cultivation of the treated embryos prior to egg transfer to the recipient females. To avoid this *ex vivo* process, we have developed a novel method called genome-editing via oviductal nucleic acids delivery (GONAD) or improved GONAD (*i*-GONAD), which enables *in situ* genome editing of zygotes present in the oviductal lumen of a pregnant female. This technology does not require any *ex vivo* handling of preimplantation embryos or preparation of recipient females and vasectomized males, all of which are often laborious and time-consuming. In this chapter, recent advances in the development of GONAD/*i*-GONAD will be described.

Keywords: *in vivo* genome editing, GONAD, *i*-GONAD, preimplantation embryos, knock out, knock-in, *in vivo* electroporation, oviducts

1. Introduction

In mammals, embryogenesis begins when the oocytes (ovulated from an ovary of a female) fertilize with spermatozoa in the uterus of the female (**Figure 1**). The fertilized oocytes, called zygotes or 1-cell stage embryos, exist at the "ampulla," an area of the oviduct located near an ovary. In mice, early zygotes are surrounded by cumulus cells (also called follicular cells) and correspond to Day 0.4 of gestation (11:00 AM in the morning after mating with a male) (box in **Figure 1**). In this case, Day 0 of gestation is defined as the day when the copulation plug is recognized in the morning. Late zygotes corresponding to Day 0.7 of gestation (16:00 PM) exhibit dissociation of cumulus cells (**Figure 1**). Then, these zygotes develop into 2-cell (~Day 1.4 of gestation), 4-cell (~Day 1.8 of gestation), 8-cell (~Day 2.4 of gestation), 16-cell (also called morula; ~Day 2.7 of gestation), early blastocyst (~Day 3.4 of gestation), and

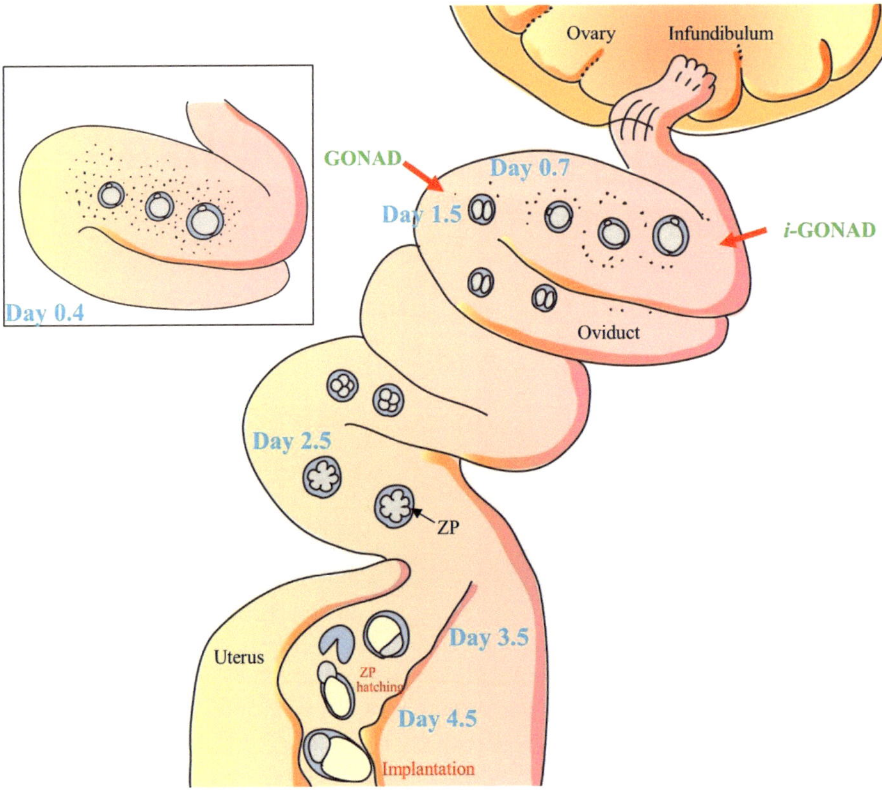

Figure 1.
Schematic of preimplantation (Days 0.5 to 4.5; Day 0 of pregnancy is defined as the day a vaginal plug is found) and postimplantation (Days 5.5) development of mice. During preimplantation, early zygote (early 1-cell embryo) (at Day 0.4; see box), late zygote (late 1-cell embryo) (Day 0.7), 2-cell embryo (day 1.5), 8- to 16-cell embryo (Day 2.5), early blastocyst (Day 3.5), and late blastocyst (Day 4.5) float in the oviductal lumen or uterine horn. Embryos at days 0.5 to 4.5 have zona pellucida (ZP), but embryos at day 4.5 begin to escape from ZP, which is called "ZP hatching," and are ready to implant into the uterine epithelium. Note that at early zygote stage (see box), fertilized eggs are tightly surrounded by cumulus cells, but at late zygote stage, cumulus cells begin to detach from an embryo. This figure was drawn in-house and reproduced with permission from Sato et al., "Recent Advances and Future Perspectives of In Vivo Targeted Delivery of Genome-Editing Reagents to Germ cells, Embryos, and Fetuses in Mice"; published by MDPI, 2020.

late blastocysts (~Day 4.4 of gestation) (at which stage implantation into the uterine epithelium starts). Notably, zygotes and 2-cell embryos exist in the ampulla, 4-cell to 8-cell embryos in the oviductal portion between the ampulla and isthmus, 16-cell embryos (morulae) in both the oviduct and uterus, and early- to late-blastocysts in the uterus (**Figure 1**).

For producing genetically engineered animals through pronuclear microinjection (MI) or viral infection, early zygotes are generally used [1]. To isolate early zygotes, oviducts dissected from pregnant female mice at ~Day 0.4 of pregnancy are teased using a needle at the ampulla under paraffin oil. The exposed cumulus-oocyte-complex is then transferred to a drop of hyaluronidase (HA)-containing medium. Brief incubation (~5 min at room temperature) relieves cumulus cells. The resulting "denuded" zygotes are always used for MI or viral infection.

Historically, the first attempt to obtain transgenic (Tg) mice was performed by microinjecting SV40 viral DNA into the blastocoel cavity of blastocysts [2]. The resulting offspring had various levels of SV40 genome in their organs. In 1980, Gordon et al. [3] first produced germ-line Tg mice through MI of exogenous DNA. Since then, successful production of Tg rabbits, sheep, and pigs was reported by Hammer et al. [4]. This MI technique relies on the physical injection of any type of nucleic acids (NAs) (i.e., purified DNA fragments of 3–4 kb carrying an expression unit) using an expensive manipulator and requires personnel with specific skill. Furthermore, it generally takes 3–4 h to finish MI using ~100 zygotes per session (**Figure 2A**). Perry et al. [5] reported a novel technique to generate Tg mice through the intracytoplasmic injection of a sperm (which has been mixed with NAs) into a zygote. Later, this technique was recognized as a useful tool to introduce large sized DNA such as bacterial artificial chromosomes (BACs) and yeast artificial chromosomes (YACs) into the mammalian genome [6–8].

Mammalian zygotes, including those of mice, are surrounded by a translucent glycoprotein layer called zona pellucida (ZP), which exists as a barrier that protects the early embryo from environmental insults, including viral infections, and injury from chemical or physical substances [9, 10]. For example, ZP-enclosed embryos could not be transduced through simple incubation with solution containing lentivirus (LV), adenovirus (AV), or retrovirus (RV) [11–13]. However, injection of those viruses into the space between ZP and zygote (called "peri-vitelline space") results in successful transduction [14, 15]. Furthermore, transduction of mouse zygotes was possible when

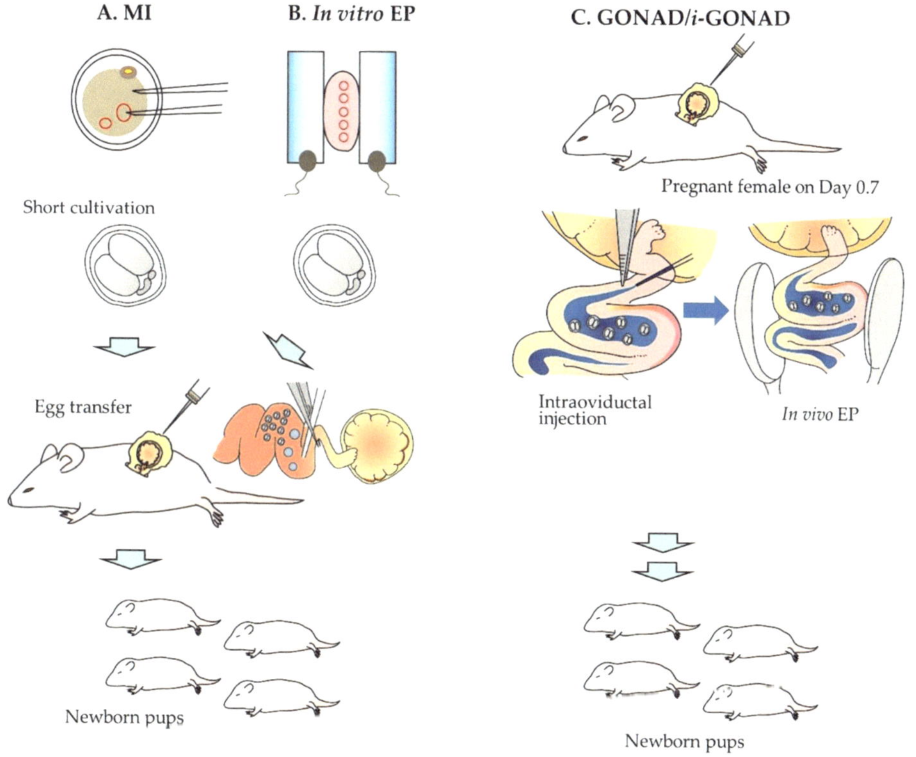

Figure 2.
Schematic of genome-edited mouse production through microinjection (MI) (A), in vitro electroporation (EP) (B), or genome-editing via oviductal nucleic acids delivery (GONAD) (or improved GONAD (i-GONAD)) (C). This figure was drawn in-house, and reproduced with permission from Sato et al., "Recent Advances in the Production of Genome-Edited Rats"; published by MDPI, 2022.

they were subjected to laser perforation of ZP in the presence of LV vectors [16]. ZP can be removed by incubating ZP-enclosed embryos in the presence of proteolytic enzyme such as pronase, or under acidic conditions using acidic Tyrode's solution [1]; therefore, gene delivery can be accomplished by incubating the ZP-removed embryos in a solution containing liposome-complexed DNA [17] or the above-mentioned viruses [11, 18]. However, these treated embryos are often vulnerable, adhesive, and are easily damaged [19]. Furthermore, transferring the ZP-removed embryos (at least up to morulae) into the oviductal lumen of pseudopregnant recipient females cannot support their normal development, because the transferred embryos tend to adhere to the oviductal epithelium [20, 21]. To avoid this, the treated embryos have to be cultured at least up to blastocysts (showing reduced adhesive property), prior to uterine transfer. Notably, acquisition of Tg founders was reported using ZP-free embryos transfected liposomally [22] or when transduced with AV vectors [11, 18].

Electroporation (EP)-mediated gene delivery is also a method that efficiently introduces exogenous NAs into a cell or zygote through electric shock-induced, transient micropores in a cell membrane [23]. It requires an expensive electroporator, but does not require a skilled person, unlike in the case of MI (**Figure 2B**). In an early study regarding EP-based introduction of DNA into mouse preimplantation embryos, DNA was first injected into the peri-vitelline space of zygotes, and then the embryos were subjected to *in vitro* EP. Unfortunately, the transfection efficiency was very low [24]. In 2002, Grabarek et al. [25] first demonstrated that *in vitro* EP enabled incorporation of small-sized NAs (i.e., siRNA) into mouse zygotes. In this case, prior to EP, ZP has to be weakened by a brief treatment with acidic Tyrode's solution to facilitate transfer of NAs into an embryo and to protect the embryos from EP damage. Notably, *in vitro* EP was also successfully used to deliver double-stranded RNA (dsRNA) and morpholinos into mouse preimplantation embryos [26, 27].

To our knowledge, Peng et al. [27] first demonstrated that plasmid DNA can be effectively delivered into mouse preimplantation embryos when they were subjected to *in vitro* EP using optimal EP parameters (i.e., voltage, pulse duration, number of pulses, and repeats). Sato et al. [28] also demonstrated that plasmid DNA can be successfully introduced into early mouse embryos present within the oviductal lumen through *in vivo* EP. *In vitro* EP-based gene delivery is generally possible using over 100 zygotes per a trial and can be finished within 15–30 min. Thus, in terms of convenience, EP appears superior to the MI-based production of transgenics. However, this success appears largely to depend on the EP parameters and the type of electroporator used, as mentioned below.

Beside *in vitro* EP, MI, and viral transduction, substances capable of penetrating ZP (also called "ZP-penetrating reagents") can be used with NAs to perform gene delivery towards ZP-enclosed embryos. For examples, Ivanova et al. [29] employed a receptor-mediated gene transfer system, with insulin as the admission ligand in the DNA-carrying construct, because early embryonic cells are known to have internalizable insulin receptors on their surface [30]. They first made an insulin-polylysine conjugated with plasmid DNA. Next, this complex was mixed with a conjugate consisting of *streptavidin*-polylysine-*biotinylated* adenovirus. Short (3 h) incubation of ZP-enclosed mouse and rabbit preimplantation embryos with the resulting complex [called "(insulin-polylysine)-DNA and (insulinpolylysine)-DNA-(streptavidin-polylysine)-(biotinylated adenovirus)"] penetrated the constructs through ZP and accumulated in the peri-nuclear space of the embryos, leading to ligand/receptor-mediated transgenesis. Joo et al. [31] developed amphiphilic *chitosan*-based nanocarriers, called VisuFect. When murine zygotes were incubated with a solution containing

Cy5.5-labeled VisuFect conjugated with poly(A) oligonucleotides, the complex gradually penetrated the cytoplasm of the ZP-enclosed zygotes. This suggests that VisuFect could be used as a vehicle to deliver NAs to ZP-intact embryos. According to Joo et al. [31], VisuFect can deliver siRNA, but not large molecules such a plasmid DNA to embryos. Nanoparticles are also a promising tool to transfer exogenous NAs into ZP-enclosed embryos. Munk et al. [32] demonstrated that multiwall carbon nanotubes (MWNTs) can cross the ZP to help the delivery of plasmid DNA (carrying the green fluorescent protein (*GFP*) gene) into bovine embryos *in vitro*. According to Munk et al. [32], MWNTs themselves are non-harmful to embryos and do not affect their viability and gene expression. On the other hand, Jin et al. [33] first demonstrated that peptide nanoparticles can introduce siRNA into an intact mouse oocyte. When oocytes were incubated with peptide nanoparticle-complexed fluorescein isothiocyanate (FITC)-conjugated siRNA for 12–14 h, a cytoplasmic fluorescence of oocytes was observed together with a target gene knockdown. Jin et al. [33] suggested that peptide nanoparticle-mediated siRNA transfection was useful to explore the function of unknown genes in mouse oocytes. Unfortunately, except for the report by Ivanova et al. [29], these studies do not show germ-line transmission or chromosomal integration of transgenes.

As mentioned above, to obtain genetically-modified (GM) animals, isolation of zygotes from pregnant females or those obtained through *in vitro* fertilization (IVF), *in vitro* gene delivery towards the isolated embryos, transient cultivation of the treated embryos, and egg transfer (ET) of the cultivated embryos to the reproductive tracts of pseudopregnant recipient females to allow the GM embryos to develop *in vivo* further. These processes are called "*ex vivo* handling of embryos" (**Figure 2A** and **B**), and generally required for MI, *in vitro* EP, liposome- or viral transduction-based transgenesis. Notably, this *ex vivo* handling of embryos is costive, labor intensive, and laborious, because it requires preparation of sterile males called "vasectomized males" to create pseudopregnant females, timely supply of those pseudopregnant females, and ET technique, which is a more difficult task requiring people with specialized skill sets. To bypass this process, direct genetic manipulation must be performed in zygotes (or embryos at more advanced stages) existing within an oviductal lumen of a pregnant female. Relloso and Esponda [34, 35] first attempted to transfect epithelial cells lining oviductal lumen by injecting liposomally encapsulated DNA directly into the oviductal lumen of a female mouse. They found that 6% of oviductal epithelial cells were successfully transfected. Rios et al. [36] also demonstrated that intraoviductal injection of naked DNA or mRNA into the estradiol-treated female rats can help incorporate those substances into the oviductal cells. The introduced DNA or mRNA will then be translated into an active protein, possibly accelerating embryo transport.

Sato [37] employed *in vivo* EP to enhance transfection efficiency in the oviductal epithelium. *In vivo* EP is a method to transfect the tissue or organ *in situ* by injecting a fluid containing plasmid DNA into the target site, holding the injection site with tweezer-type electrodes, and subsequently giving an electric shock using an electroporator [38]. Using this system, several organs/tissues including kidney [39, 40], liver [41], brain [42], skin [43, 44], skeletal muscle [45], testis [46, 47], efferent duct [48], ovary [49], and fetuses [50–52], have been successfully transfected. According to Sato [37], ovary/oviduct/uterus were pulled out and exposed on the back of a female on Day 0.4 of pregnancy (~11:00 AM; corresponding to early zygotes). Then, a small amount (1–2 μL) of solution which contains an enhanced green fluorescent protein (eGFP) expression plasmid (0.5 μg/μL) and 0.2% (v/v) trypan blue (TB) (used as a marker for visualizing injected materials) was injected into the oviductal lumen using a mouthpiece-controlled glass pipette under a dissecting microscope. Immediately after injection, an entire oviduct was subjected to *in vivo* EP using tweezer-type electrodes in an attempt to transfect oviductal epithelium facing oviductal lumen and possibly zygotes (floating in the oviductal lumen) with the exogenous DNA. The EP condition was eight square-wave pulses with a pulse duration of 5 ms and an electric field intensity of 50 V, generated by a square-wave pulse generator (#T-820; BTX Genetronics Inc., San Diego, CA, USA). After EP, the treated oviduct was returned to the original position. One day after the surgery, oviducts were dissected from the female to isolate 2-cell embryos. When the eGFP-derived fluorescence in isolated embryos and oviducts was inspected, maximal 43% of oviductal epithelial cells facing oviductal lumen were fluorescent, while no fluorescence was discernible in the isolated embryos; only a cellular remnant probably derived from a part of zygotes was found to be fluorescent. Based on this finding, Sato [37] speculated that failure of gene delivery to zygotes may be due to the cumulus cells surrounding zygotes acting as a barrier. Because the oviductal epithelium can be efficiently transfected using *in vivo* EP method, Sato [37] named this technology "gene transfer via oviductal epithelium (GTOVE)."

Sato et al. [28] next attempted to transfect 2-cell embryos floating in the oviductal lumen using GTOVE, since those embryos are already free of cumulus cells. To address this issue, GTOVE was performed in pregnant females at Day 1.4 of pregnancy using the same conditions elaborated in the study by Sato [37]. One day after the GTOVE procedure, 8-cell embryos were collected from the GTOVE-treated females for checking eGFP-derived fluorescence. Of the 12 oviducts (6 females used) examined, 3 contained fluorescent 8-cell stage embryos (33%, 19/58 tested), but the intensity of fluorescence varied among the embryos. Unfortunately, gene expression was transient in this system, with no evidence for chromosomal integration of transgenes [28]. These results indicate that successful *in vivo* introduction of exogenous plasmid DNA into early mouse embryos is possible, as far as the T-820 electoporator is employed. However, the T-820 electoporator is currently unavailable. When we performed GTOVE to transfect 2-cell embryos using another electroporator (NEPA21; NEPA GENE Co., Chiba, Japan), the collected embryos failed to fluoresce [53]. This suggests a need to carefully examine the optimal EP condition enabling ZP penetration of larger sized molecules like plasmid DNA, as suggested by Peng et al. [27] and Hakim et al. [54].

2. Development of genome editing technology

Genome-editing technology includes zinc-finger nucleases (ZFNs), transcription activator-like effector nucleases (TALENs), and clustered regularly interspaced short palindromic repeats (CRISPR)/CRISPR-associated protein 9 (CRISPR/Cas9), all of which employ sequence-specific nucleases to induce modifications in a predefined region of the genome (reviewed by Harrison et al. [55]; Hsu et al. [56]). In the absence of the donor (or template) DNA [including longer genes (>1 kb), single-stranded (ss) sequences (>200 bp) or synthetic oligodeoxynucleotides (ODNs) 20–30 bp in size, all of which have sequences showing homology to the target sequence, these nucleases can induce double-strand breaks (DSBs) which are then repaired *via* nonhomologous end joining (NHEJ). This repair occurs through the cellular machinery, which frequently generates random insertions, deletions, or substitutions of nucleotides (called indels) at the break site. These indels often cause frameshift mutations, leading to the occasional failure in protein expression. In the presence of the donor (or template) DNA, it will be introduced into the DSB site through homology-directed repair (HDR), and this event is called "knock-in (KI)." According to Yoshimi et al. [57], KI is more difficult to complete successfully than inducing NHEJ-mediated indels. Furthermore, NHEJ occurs in nondividing and dividing cells, but HDR occurs preferentially in dividing cells [58, 59].

Unlike ZFN and TALEN, the CRISPR/Cas9 gene editing system has several advantages, including easy design for any genomic targets, simplicity, and the ability to modify several target genes simultaneously (multiplexing). Owing to these properties, the CRISPR/Cas9 system is now widely used in various biological systems. It employs only two components: (1) a guide RNA (gRNA), comprised either of a duplex CRISPR RNA (crRNA)/trans-activating CRISPR RNA (tracrRNA) molecule or of single-guide RNA (sgRNA), a fusion between crRNA and tracrRNA, and (2) a Cas9 endonuclease (reviewed by Harrison et al. [55]; Hsu et al. [56]). The gRNA can bind to the specific DNA sequence together with Cas9. Once bound, the Cas9 cleaves dsDNA 3 bp upstream of the protospacer adjacent motif (PAM, 5′-NGG-3′), which is recognized by the Cas9 protein. The cleaved site is then repaired by various cellular machineries, such as NHEJ and HDR.

3. Development of novel technologies enabling EP-based genome-editing *in vitro* and *in vivo*

As described in Section 2, in the late 2013, CRISPR/Cas9 system was recognized as a useful tool to manipulate a target gene in mammalian cells and embryos (reviewed by Harrison et al. [55]; Hsu et al. [56]). Since then, genome-edited animals from various species including mice, rats, pigs, bovines, and primates (monkeys) have been generated through MI [60–66]. Kaneko et al. [67] first demonstrated that *in vitro* EP of rat zygotes in the presence of genome editing reagents is a powerful tool to produce GM animals (**Figure 2B**). According to Kaneko et al. [67], intact rat zygotes were electroporated using an NEPA21 electroporator in a solution containing ZFN (40 μg/mL) mRNA [targeted to the rat interleukin 2 receptor subunit gamma gene (*Il2rg*)] under the EP condition of a poring pulse (Pp) (voltage: 225 V; pulse interval: 50 ms; pulse width: 1.5 and 2.5 ms; number of pulses: 4). The mechanism underlying gene delivery by this system is described in our previous paper [68]. As a result, they obtained genome-edited rat offspring with an efficiency of 73%, which is roughly 2-fold higher than that obtained through MI-mediated genome editing. Notably, they confirmed germ-line transmission of the genome-edited traits beyond next generation. This technology was thus named "technique for animal knockout system by electroporation (TAKE)." Since then, various genome-edited animals including mice [69–75], rats [76–78], and pigs [79–81] have been successfully generated using this technology. Furthermore, this technology has been applied to introduce nucleases into frozen embryos [82] and embryos derived from freeze-dried and frozen sperm [83].

Takahashi et al. [84] developed a novel *in vivo* EP-based method enabling *in situ* CRISPR/Cas9-based genome editing towards early mouse embryos (2-cell embryos) using a technique similar to GTOVE (**Figure 2C**). In this case, a solution (1–1.5 μL) containing *Cas9* mRNA (up to 1.1 μg/μL), sgRNA (0.6 μg/μL; targeted to *eGFP* cDNA), and 0.2% (v/v) TB was injected through the oviductal wall under observation into the oviductal lumen of pregnant non-Tg females at Day 1.4 of pregnancy (corresponding to the 2-cell stage) that had been mated with Tg males containing an *eGFP* expression cassette in a homozygous manner [85]. In this case, the pregnant female should have zygotes, all of which are expected to have *eGFP* expression cassette in a heterozygous manner, and be fluorescent (**Figure 3**). If CRISPR/Cas9-induced mutations (indels) occur in the genomes of those zygotes (*eGFP* cDNA), some of their offspring will most likely lose fluorescence from their entire body. Fluorescence inspection of mid-gestational fetuses dissected from the *in vivo* EP-treated females demonstrated that out of 6 fetuses obtained, two lost fluorescence completely, two exhibited weak fluorescence, and fluorescence in the remaining two fetuses remained unaltered. Molecular biological analysis revealed that the former fetuses comprised knock out (KO) cells, the middle fetuses contained a mixture of KO cells and intact cells (whose state is called "mosaic mutation"), and the latter fetuses had unedited intact cells. Based on these findings, the KO efficiency of these fetuses

Figure 3.
Experimental flowchart for genome-editing via oviductal nucleic acids delivery (GONAD). Females (C57BL/6) are first mated to C57BL/6-Tg (CAG-eGFP) male mice that possess EGFP transgenes in a homozygous (Tg/Tg) state. All the fetuses are then expected to be eGFP-expressing fetuses carrying the transgenes in a heterozygous (Tg/+) state. Thus, successful genome editing targeted to eGFP at preimplantation stages are expected to reduce the levels of eGFP fluorescence in the mid-gestational fetuses, as a result of genome editing in the chromosomally integrated eGFP transgenes. When GONAD was performed using Cas9 mRNA and sgRNA towards pregnant females at Day 1.4 of pregnancy, there were three types of fetuses (showing complete loss of fluorescence, partial fluorescence, or no reduction in fluorescence) in view of fluorescence expression pattern.

can be estimated to be approximately 29%. Based on these findings, Takahashi et al. [84] named this technology "genome-editing via oviductal nucleic acids delivery (GONAD)."

4. Development of improved GONAD (*i*-GONAD)

In 2018, Ohtsuka et al. [86] in the same group of Takahashi et al. [84] further elaborated on the GONAD technology. As was mentioned in Section 3, GONAD enables *in situ* genome editing in 2-cell embryos. In this case, only one blastomere among two blastomeres can be genome-edited, generating "mosaic" fetuses comprising edited and non-edited cells. To avoid this risk, genome editing at zygote (one-cell) stage is desirable. As was mentioned previously, early zygotes (corresponding to Day 0.4 of pregnancy; 11:00 AM) are tightly surrounded by cumulus cells. In our previous experience, GONAD at this stage failed, because almost all of the genome editing reagents introduced intaoviductally was trapped by the cumulus cells [86]. On the other hand, as mentioned previously, the detachment of cumulus cells from a zygote commences at late zygote stage (corresponding to Day 0.7 of pregnancy; 16:00 PM). Based on this finding, Ohtsuka et al. [86] first attempted to disrupt the forkhead box protein E3 (*Foxe3*) locus using ribonucleoprotein (RNP) (comprised of 1 µg/µL of Cas9 protein and 30 µM of crRNA/tracrRNA) using pregnant females at Day 0.7 of pregnancy. The advantage of using RNP is to induce genome editing more rapidly than using *Cas9* mRNA [87]. *In vivo* EP was performed using the NEPA21 apparatus under the following conditions: Pp: 50 V, 5-ms pulse, 50-ms pulse interval, three pulses, 10% decay (± pulse orientation); Tp: 10 V, 50 ms pulse, 50 ms pulse interval, three pulses, and 40% decay (± pulse orientation). This modification resulted in 97% of the embryos exhibiting indels in the target locus. They next attempted to perform KI of a sequence (coding for a gene of interest (GOI)) into the target locus using RNP containing 1–2 µg/µL of ssODN or 0.85–1.4 µg/µL of ssDNA (with ~925 bases in size) generated through a novel method, called Easi-CRISPR, a highly efficient KI technique using ssDNA as donor templates [88, 89]. As a result, ~50% and ~ 15% of embryos were found to have KI alleles for ssODN and longer ssDNA in their genome, respectively.

Ohtsuka et al. [86] also demonstrated that large deletion (LD) of a target sequence can be accomplished using this modified GONAD, which was re-named as "*i*-GONAD." It uses the Cas9 protein instead of the *Cas9* mRNA, and targets late zygotes. For example, they designed two gRNAs (16.2 kb distance apart), both of which recognize either the sites of the retrotransposon sequence in the intron 1 of *Agouti* locus in the C57BL/6 mouse genome. The *i*-GONAD-mediated deletion of the inserted sequence resulted in the generation of fetuses with agouti coat color with efficiencies of 50%. Molecular biological analysis of these rescued offspring revealed the evidence for LD in the *Agouti* locus containing retrotransposon of C57BL/6 mice.

Since the studies by Takahashi et al. [84] and Ohtsuka et al. [86], several reports have been provided using GONAD/*i*-GONAD technologies. In **Table 1**, past studies using those technologies are listed. Also, the detailed protocols for *i*-GONAD in mice have been provided by Gurumurthy et al. [116, 117] and Ohtsuka and Sato [118]. The GONAD/*i*-GONAD-based production of genome-edited rats is also possible using the same approach shown in mice. Notably, Namba et al. [119] demonstrated the protocols for GONAD/*i*-GONAD in rats.

Type of method (content of CRISPR/Cas9 reagents)	Genome editing tool (mode for gene modification) EP apparatus	Animal (species strain)	Outcome	Target gene	References
GONAD (mRNA)	CRISPR/Cas9 (indels) BTX T820	Mouse (MCH(ICR), eGFP Tg)	Performed at Day 1.5 of pregnancy (corresponding to 2-cell stage) in mice; first successful genome editing at mid-gestational fetuses with 28% efficiency; also frequently associated with mosaic mutations	*Hprt*, *eEF2*, *eGFP*	Takahashi et al. [84]
i-GONAD (mRNA, protein)	CRISPR/Cas9, Cas12a (indels, KI) BTX T820, NEPA21, CUY21EDIT II	Mouse (MCH(ICR), C57BL/6, BALB/cA, C3H/He, DBA2, B6D2F1)	Performed at Day 0.7 of pregnancy (corresponding to late zygote stage); successful genome editing in offspring with efficiencies of 50 ~ 100% for indels, ~50% for ssODN KI (single-base changes) and 15% for longer ssDNA KI; kilobase-sized deletions can also be induced	*Foxe3*, *Pitx3*, *Tis21* *Tyr*, *Agouti* *Kit*, *Cdkn1a*, *Cdkn2a*	Ohtsuka et al. [86]
rGONAD (protein)	CRISPR/Cas9 (indels, KI) NEPA21	Rat (DA, WKY)	Showing successful genome editing (with efficiencies of 50.0% and 17.8% for indels and 26.9% and 11.1% for KI) using *i*-GONAD in rats; the *i*-GONAD in rats was renamed as "rGONAD."	*Tyr*	Kobayashi et al. [90]
i-GONAD (protein)	CRISPR/Cas9 (indels, KI) NEPA21	Rat (SD, LEW, SD x BN)	Showing successful genome editing (with efficiencies of ~62% for indels and ~ 9% for KI) using *i*-GONAD in various rat strains; abnormal facial morphogenesis in fetuses was induced when *Pax6* locus was targeted; strain-difference regarding the optimal *in vivo* EP condition was noted for BN strain	*Tyr*, *Pax6*	Takabayashi et al. [91]

Type of method (content of CRISPR/Cas9 reagents)	**Genome editing tool (mode for gene modification) EP apparatus**	**Animal (species strain)**	**Outcome**	**Target gene**	**References**
AAV-based GONAD	CRISPR/Cas9 spCas9 (indels)	Mouse (C57BL/6)	Intraoviductal injection of a solution containing two rAAV serotype 6 vectors, one carrying *spCas9* gene and the other carrying gRNA expression unit, targeted *Tyr* into pregnant female mice at Day 0.5 of pregnancy; led to production of genome-edited pups with 10% efficiency	*Tyr*	Yoon et al. [92]
i-GONAD(protein)	CRISPR/Cas9 (chromosome inversion, LD) NEPA21	Mouse (C57BL/6, B6C3F1)	Showing the first successful target-specific chromosomal inversions of 7.67 megabases (Mb) in length in mice; this is longer than any previously reported inversion produced using PI-based methods.	*Pafah1b1, Gm30413, Rad51*	Iwata et al. [93]
i-GONAD (protein)	CRISPR/Cas9 (indels) NEPA21	Golden (Syrian) hamster	A review showing that *i*-GONAD is useful for production of *Tyr* KO hamsters; the hamsters showed albino coat color, as expected (Hirose and Ogura, unpublished).	*Tyr*	Hirose and Ogura [94]
i-GONAD (protein)	CRISPR/Cas9 (KI) NEPA21	Mouse (C57BL/6)	Successful production of *p21* KO mice with exacerbation of fibrosis.	*Cdkn1a*	Koyano et al. [95]
i-GONAD (protein)	CRISPR/Cas9 (indels) NEPA21	Mouse (B6C3F1)	Successful generation of early embryos and fetuses with complete loss of α-Gal epitope expression	*GGTA1*	Sato et al. [53]
i-GONAD (protein)	CRISPR/Cas9 (indels) NEPA21	Golden (Syrian) hamster	Showing successful production of *acrosin* KO hamsters; homozygous mutant males were completely sterile.	*Acrosin*	Hirose et al. [96]

Type of method (content of CRISPR/Cas9 reagents)	Genome editing tool (mode for gene modification) EP apparatus	Animal (species strain)	Outcome	Target gene	References
Sequential *i*-GONAD (protein)	CRISPR/Cas9 (indels, KI) NEPA21	Mouse (B6C3F1)	Showing the usefulness of two steps of *i*-GONAD (at Day 0.7 and Day 1.4–1.5 of pregnancy) to induce two mutations which are closely located each other; can create floxed mice carrying two *lox* sites flanking an exon; this approach is named "sequential *i*-GONAD (*si*-GONAD)"	*GGTA1*, *Mecp2*	Sato et al. [97]
i-GONAD (protein)	CRISPR/Cas9 (indels, KI) NEPA21, GEB15	Mouse (C57BL/6, BALB/c, ICR)	For obtaining genome-edited C57BL/6 mice, setting a constant current of 100 mA upon *in vivo* EP is recommended; in this study, optimal EP conditions allowing the generation of a 100 mA current using two electroporators, NEPA21 and GEB15, are explored; consequently, the current and resistance were set to 40 V and 350–400 W, respectively, and were found to be suitable for *i*-GONAD using C57BL/6 mice.	*Tyr*	Kobayashi et al. [98]
i-GONAD (protein)	CRISPR/Cas9 (KI, LD) NEPA21, CUY21EDIT II	Rat (BN)	Explored for optimal condition for obtaining genome-edited BN rats through *i*-GONAD; under a current of 100–300 mA using NEPA21, genome-edited BN rats were obtained with efficiencies of 75–100%; under a current of 150–200 mA using CUY21EDIT II genome-edited BN rats were obtained with efficiencies of 24–55%.	*Tyr*	Takabayashi et al. [99]

Type of method (content of CRISPR/Cas9 reagents)	Genome editing tool (mode for gene modification) EP apparatus	Animal (species strain)	Outcome	Target gene	References
i-GONAD (protein)	CRISPR/Cas9 (indels) NEPA21	Mouse (ICR)	Showing the usefulness of pretreatment with hyaluronidase before *i*-GONAD at Day 0.4 of pregnancy (corresponding to early zygote stage).	*Fgf10*	Kaneko and Tanaka [100]
i-GONAD (protein)	CRISPR/Cas9 (indels) CUY21EDIT II	Mouse (CD-1, C57BL/6)	Showing successful production of floxed alleles for five genes with an efficiency of 10% through a single step under relatively low costs and minimal equipment setup; constitutive KO alleles were obtained as byproducts of these experiments.	*Fosl1, Plagl1, Ak040954, Clcf1, Gm44386*	Shang et al. [101]
i-GONAD (protein)	CRISPR/Cas9 (indels) ?	Mouse (C57BL/6)	Showing successful production of *Klrc2* KO mice; Qa-1b expression levels were down-regulated in infected cells but increased in some bystander immune cells to respectively promote or inhibit their killing by activated natural killer cells.	*Klrc2*	Ferez et al. [102]
i-GONAD (protein)	CRISPR/Cas9 (KI) NEPA21	Mouse (ICR, C57BL/6)	Showing successful production of hemagglutinin (HA)-tag KI mice; HA-tag sequence was successfully inserted into the C terminus of the *ATF5* coding sequence.	*ATF5*	Nakano et al. [103]
i-GONAD (protein)	CRISPR/Cas9 (indels) NEPA21	Mouse (ICR)	Showing successful production of *Gbx2* KO mice showing lack of thalamocortical axons.	*Gbx2*	Yoshinaga et al. [104]
i-GONAD (protein)	CRISPR/Cas9 (indels) ?	Mouse (C57BL/6)	Showing successful production of *Serpina3* KO mice which failed to evoke proper resolution, indicating that *Serpina3n* has a physiological function in resolving inflammation.	*Serpina3n*	Ho et al. [105]

Type of method (content of CRISPR/Cas9 reagents)	Genome editing tool (mode for gene modification) EP apparatus	Animal (species strain)	Outcome	Target gene	References
i-GONAD (protein)	CRISPR/Cas9 (indels) ?	Mouse (C57BL/6)	Showing successful production of *Nrsn2* KO mice with reduced AMPAR signaling; Neurensin-2 was found to have a role as a novel modulator of emotional behavior.	*Nrsn2*	Umschweif et al. [106]
i-GONAD (protein)	CRISPR/Cas9 (indels) NEPA21	Mouse (C3H/He, C57BL/6)	Showing the first *Tprkb* KO mouse with an embryonic lethal mutation that was stably maintained in heterozygotes as inversion balancer strains using a B6.C3H-In(6)1J inversion identified from C3H/HeJJcl.	*Tprkb*	Iwata et al. [107]
i-GONAD (protein)	CRISPR/Cas9 (indels) NEPA21	Golden (Syrian) hamster	Showing successful production of *Mov10l1* KO hamster which is sterile in both sexes; this is in contrast with the case of *Mov10l1* KO mice which were known to be sterile in males, but not in females.	*Mov10l1*	Loubalova et al. [108]
i-GONAD (protein)	CRISPR/Cas9, Cas12a (indels, KI) CUY21Editll	Mouse (△eGFP Tg)	Using △eGFP Tg mice carrying a single copy of disrupted *eGFP* (△eGFP), the feasibility of *i*-GONAD-mediated gene correction using AsCas12a nuclease is shown.	*eGFP*	Miura et al. [109]
i-GONAD (protein)	CRISPR/Cas9 (intronic deletion) CUY21Editll	Mouse (ICR)	Showing successful production of *Dll1*Δ232 mouse model in which E box motifs from intron 4 of *Dll1* gene have been deleted.	*Dll1*	Zhang et al. [110]
rGONAD (protein)	CRISPR/Cas9 (indels) NEPA21, GEB15	Rat (WKY)	Showing successful production of *Col4α5* KO rat model for Alport syndrome with hematuria, proteinuria, and high levels of BUN/Cre; died at 18 to 28 weeks of age.	*COL4A3, COL4A4, COL4A5*	Namba et al. [111]

Type of method (content of CRISPR/Cas9 reagents)	**Genome editing tool (mode for gene modification) EP apparatus**	**Animal (species strain)**	**Outcome**	**Target gene**	**References**
i-GONAD (protein)	CRISPR/Cas9 (indels) NEPA21, CUY21EDIT II	Rat (SD)	To improve the efficiency of *i*-GONAD-mediated KI in rats, three gRNAs (crRNA1, crRNA2, and crRNA3), all of which recognize the target sites that are located very closely each other, were tested; consequently, KI efficiency varied among those gRNAs, suggesting that the choice of gRNA is important for determining the KI efficiency; the use of KI-enhancing drugs failed to increase the KI efficiency.	*Tyr*	Aoshima et al. [112]
i-GONAD (protein)	CRISPR/Cas9 (KI) NEPA21	Mouse (C57BL/6)	Showing successful production of RDEB mouse model with *COL7A1* mutations, c.5818delC and E2857X; 5818delC homozygous mice developed severe RDEB-like phenotypes and died immediately after birth, whereas E2857X homozygous mice did not have a shortened lifespan compared to WT mice; adult E2857X homozygous mice showed hair abnormalities, syndactyly, and nail dystrophy.	*COL7A1*	Takaki et al. [113]
i-GONAD (protein)	CRISPR/Cas9 (KI) NEPA21	Mouse (C57BL/6)	Showing successful production of *Axdnd1* KO mice showing sterility caused by impaired spermiogenesis.	*Axdnd1*	Hiradate et al. [114]

Type of method (content of CRISPR/Cas9 reagents)	Genome editing tool (mode for gene modification) EP apparatus	Animal (species strain)	Outcome	Target gene	References
i-GONAD (protein)	CRISPR/Cas9 (indels) CUY21EditII	Mouse (C57BL/6)	Effects of AIMA treatment on an increased number of littermate in C57BL/6 mice were examined; the mean litter size following *i*-GONAD increased from 4.8 to 7.3 after the AIMA treatment; genetic modifications were confirmed in 80/88 (91%) of the offspring.	*Tyr*	Hasegawa et al. [115]

Abbreviations: AAV, adeno-associated virus; Agouti, Agouti-signaling protein (ASIP); AIMAs, anti-inhibin monoclonal antibodies; AMPAR, α-amino-3-hydroxy-5-methyl-4-isoxazole propionic acid receptor; ATF5, activating transcription factor 5; Axdnd1, axonemal dynein light chain domain containing 1; B6C3F1, a cross between female C57BL/6 and male C3H/He mice; B6D2F1, a cross between female C57BL/6 and male DBA/2 mice; BN, Brown Norway rat; Cas12a (AsCas12a), class 2 CRISPR-Cas endonuclease Cas12a (previously known as Cpf1); CCK, cholecystokinin; Cdkn1a, cyclin dependent kinase inhibitor 1A; Cdkn2a, cyclin dependent kinase inhibitor 2A; Clcf1, cardiotrophin-like cytokine factor 1; COL4A3, collagen type IV α3 chain; COL4A4, collagen type IV α4 chain; COL4A5, collagen type IV α5 chain; COL7A1, collagen type VII α1 chain; CRISPR/Cas9, clustered regularly interspaced short palindromic repeats (CRISPR)/CRISPR-associated protein 9; DA, Dark Agouti rat; Dll1, Delta like canonical Notch ligand 1; eEF2, eukaryotic translation elongation factor 2; eGFP, enhanced green fluorescent protein; EP, electroporation; Fgf10, fibroblast growth factor 10; Fosl1, Fos-like antigen 1; Foxe3, forkhead box E3; Gbx2, gastrulation brain homeobox 2; GGTA1, α-1,3-galactosyltransferase; GONAD, genome-editing via oviductal nucleic acids delivery; Hprt, hypoxanthine guanine phosphoribosyl transferase; i-GONAD, improved GONAD; indels, insertions, deletions, or substitutions of nucleotides; KI, Knock-in; Kit, KIT proto-oncogene, receptor tyrosine kinase; Klrc2, killer cell lectin like receptor C2; KO, Knock out; LD, large deletion; LEW, Lewis rat; Mecp2, methyl-CpG binding protein 2; Mov10l1, Mov10 like RISC complex RNA helicase 1; Nrsn2, Neurensin 2; Pafah1b1, platelet activating factor acetylhydrolase 1b regulatory subunit 1; Pax6, paired box 6; Pitx3, paired like homeodomain 3; Plagl1, pleiomorphic adenoma gene-like 1; Qa-1b, a 48 kDa non-classical MHC class Ib molecule; rAAV, recombinant adeno-associated viruses; Rad51, RAD51 recombinase; RDEB, recessive dystrophic epidermolysis bullosa; SD, Sprague–Dawley rat; Serpina3n, serpin family A member 3; si-GONAD, sequential i-GONAD; spCas9, Streptococcus pyogenes-derived Cas9; ssODN, single-stranded oligodeoxynucleotide; Tis21, TPA-inducible sequences 21; Tg, transgenics; Tprkb, TP53RK binding protein; Tyr, tyrosinase; WKY, Wistar Kyoto rat.

Table 1.
Summary of genome-edited animals produced using GONAD/i-GONAD between the years 2015–2020.

5. GONAD/*i*-GONAD in rats

Rats (*Rattus norvegicus*) and mice (*Mus musculus*) are both classified into the same rodent family, and have been recognized as the most widely used models for biomedical research during the past four decades. However, these animals are different on many factors. For example, the rat is larger (roughly about eight- to ten-fold) in size than the mouse. Therefore, rats have been extensively used for pharmacological and surgical research, as exemplified by easier and more rapid microsurgery, multiple sampling of blood and tissues with relatively large amounts, and precise injection of substances into blood vessels (reviewed by Kjell and Olson [120]). Additionally, rats are considered as animals suitable for toxicological, neurobehavioral, and cardiovascular studies (reviewed by Jacob [121]). Since the first report in 1997 by Guerts et al. [122] on the production of genome-edited rats using ZFN technology, a total of 113 GM rats by MI (including pronuclear MI and cytoplasmic MI) or 9 rats by *in vitro* EP method have been produced during the period 1997–2021 (reviewed by Sato et al. [68]). Notably, GM rats often exhibited disease phenotype similar to that observed in humans, comparable to those shown in GM mice. For example, Zhang et al. [123] produced KO rats through MI of CRISPR/Cas9 components to obtain rat models for hereditary tyrosinemia type I (HT1), a disease caused by a deficiency in fumarylacetoacetate hydrolase (FAH) enzyme. These *Fah* KO rats developed remarkable liver fibrosis and cirrhosis, which have not been observed in *Fah* mutant mice. Furthermore, dystrophin-coding gene (dystrophin) (*Dmd*) KO rats (but not mice) presented cardiovascular alterations close to those observed in humans, which are the main cause of death in patients [124, 125]. These findings encourage the speculation that rats may better mimic the human situation than mice.

In 2018, two groups in Japan succeeded in generating genome-edited rats using *i*-GONAD. Kobayashi et al. [90] examined optimal *in vivo* EP conditions, allowing successful *i*-GONAD using NEPA21 electroporator and red fluorescent dextran (RFD) (tetramethylrhodamine-labeled dextran 3 kDa) as a fluorescent dye to monitor the fate of injected materials. Wistar Kyoto (WKY) strain rats were first subjected to the intraoviductal injection of a solution containing RFD + TB into pregnant female rats at Day 0.7 of pregnancy and subsequent *in vivo* EP using the NEPA21 electroporator under the following EP conditions: Pp: 30, 40, or 50 V, 5-ms pulse, 50-ms pulse interval, 3 pulses, and 10% decay (± pulse orientation); Tp: 10 V, 50-ms pulse, 50-ms pulse, 6 pulses, and 40% decay (± pulse orientation). When two-cell embryos recovered one day after *i*-GONAD were inspected for RFD-derived fluorescence, EP with 50 V for Pp and 6 times for Tp yielded maximal fluorescence in those embryos, with 74% efficiency. Using these optimal EP conditions, they attempted to disrupt the endogenous tyrosinase (*Tyr*) gene, a gene coding for protein essential for eye pigmentation, in pigmented females through intraoviductal injection of RNP (1 μg/μL of Cas9 protein +30 μM of sgRNA targeted to *Tyr*). As a result, genome-edited pups with albino-colored coat were obtained with an efficiency of 42%, when a Pp of 50 V was employed. They also attempted to recover the coat-color mutation in WKY females using an ssODN-based KI approach. Consequently, the KI efficiency was 27% in the pups born. They named this rat-based *i*-GONAD as "rGONAD."

Takabayashi et al. [91] provided data similar to that of Kobayashi et al. [90], demonstrating that *i*-GONAD-based mutations (indels) resulted in the generation of fetuses (derived from pigmented Brown Norway (BN) × albino SD rat crosses) with non-pigmented eyes, with an efficiency of 56%. In this case, the *in vivo* EP condition used was almost the same as that used for *i*-GONAD in mice. They also tested the possibility of KI (targeted *Tyr* locus) using albino Lewis (LEW) rats and demonstrated that the *i*-GONAD-mediated KI efficiency was as low as ~5%, when the presence of fetuses with pigmented eyes was assessed. Takabayashi et al. [91] further attempted to disrupt another endogenous gene, paired box 6 (*Pax6*), an essential locus required for facial development, using *i*-GONAD. Out of 8 mid-gestational fetuses obtained, three had completely lacked eyes and lateral nasal prominence.

6. GONAD/*i*-GONAD in hamsters

The golden hamster (*Mesocricetus auratus*) is a small rodent that has been extensively used in biomedical research in fields including oncology, immunology, metabolic disease, cardiovascular disease, infectious disease, physiology, and behavioral and reproductive biology [126]. In 1976, hamster oocytes were first used for IVF assay to test the fertilizing ability of human spermatozoa [127]. However, hamster embryos are highly vulnerable to *in vitro* conditions, hindering the generation of GM hamsters [128].

Hirose et al. [96] attempted to produce GM hamsters using *i*-GONAD, which allows embryo manipulation under the environment where the effects of handling embryos *in vitro* can be avoided as possible. Through intraoviductal injection of six sgRNAs (targeted *acrosin* gene) and the Cas9 protein, they produced KO hamsters lacking expression of acrosin, a protein thought to be essential for sperm penetration through ZP, to investigate how acrosin-KO hamster spermatozoa behaved both *in vivo* and *in vitro.* A total of 15 pups obtained, eight of which were weaned. Of these, five were found to have mutant alleles. Homozygous mutant males were completely sterile, as the mutant spermatozoa attached to ZP, but failed to penetrate it. This finding indicates that in hamsters, acrosin plays an indispensable role in allowing fertilizing spermatozoa to penetrate ZP.

7. Application of GONAD/*i*-GONAD

7.1 Two-step *i*-GONAD can introduce mutations at two sites located close to each other

CRISPR/Cas9-based introduction of indels into two sites (which are closely located each other) through a single shot of transfection has been recognized difficult, because frequent deletion between these two sites occurs. However, sequential transfection of genome editing reagents can avoid such deletion between the two sites. To test the possibility, Sato et al. [97] tried to generate two types of indels at two target sites (that are located very close to each other; 44 bp apart) by performing *i*-GONAD sequentially (**Figure 4**). The two gRNAs were first designed to recognize the upper and lower portion of exon 4 of α-1,3-galactosyltransferase gene (*GGTA1*), coding for the protein essential for synthesizing the cell-surface α-Gal epitope [129]. For the 1st *i*-GONAD, a solution containing Cas9 protein, sgRNA (termed "A") recognizing the upper portion of exon 4, and dye (Fast Green FCF; for monitoring injection process) was injected intraoviductally at Day 0.7 of pregnancy (corresponding to late zygote stage). Next day, a solution containing Cas9 protein, sgRNA recognizing the lower portion of exon 4 (termed "B"), and dye was injected intraoviductally at Day 1.7–1.8 of pregnancy (corresponding to 2-cell stage). One day after the final surgery, morulae were isolated for single embryo-based analysis for possible indels at the target sites. As a result, the efficiency of successful generation of morulae with indels at both two sites was 18%. In contrast, *i*-GONAD using two sgRNAs (A and B) + Cas9 protein at Day 0.7 of pregnancy failed to generate morulae with mutations in both sites at exon 4 of *GGTA1*. Based on these findings, Sato et al. [97] named this approach "sequential *i*-GONAD (*si*-GONAD)."

7.2 Preparation of floxed mice using *i*-GONAD

The Cre/*loxP* system is a useful tool for assessing *in vivo* gene function. Spatially and temporally-controlled expression of Cre recombinase enables precise deletion of *loxP*-floxed chromosomally integrated GOI. To realize it, two *loxP* sites must be simultaneously inserted in *cis* into the target locus. The resulting mice are called "conditional KO mice." Previously, this process for conditional KO mouse production was achieved by embryonic stem (ES) cell-based gene targeting and subsequent chimeric

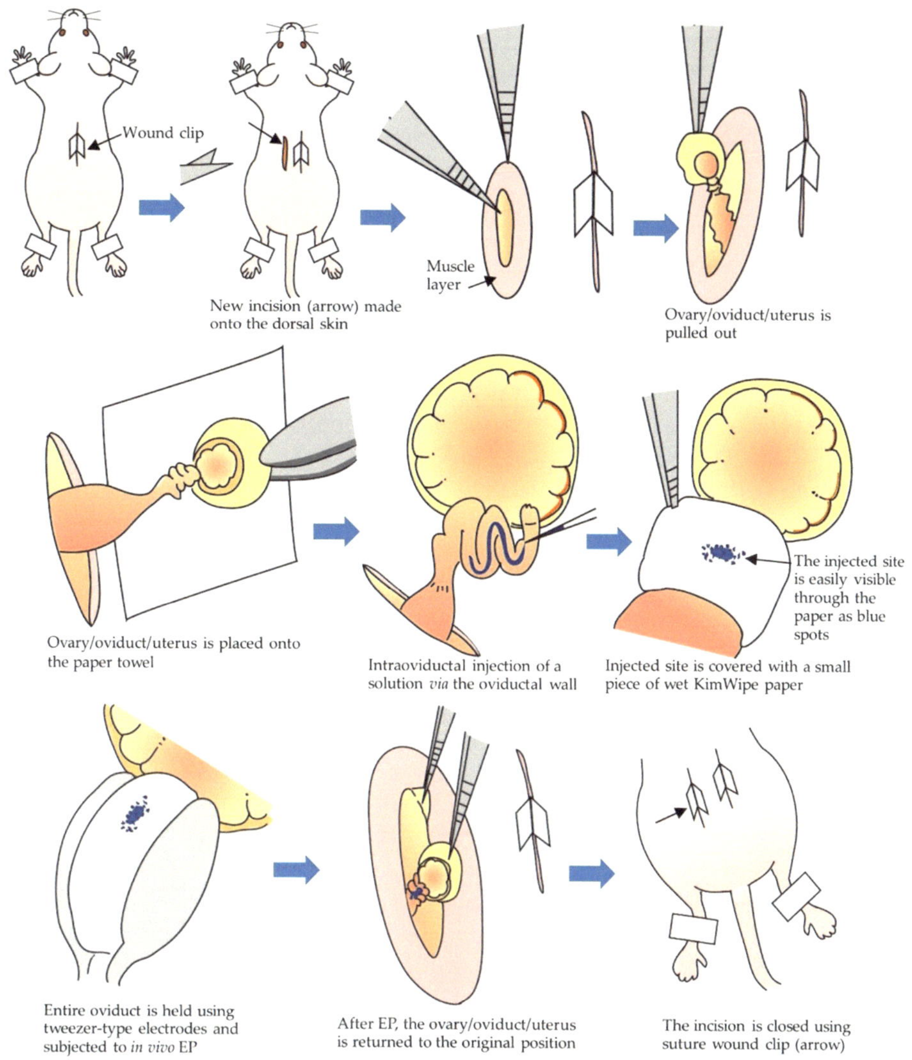

Figure 4.
Schematic of the detailed procedure of sequential improved genome-editing via oviductal nucleic acid delivery (si-GONAD). After first i-GONAD on Day 0.7 of pregnancy, second i-GONAD is performed as shown in the panel on Day 1.7 of pregnancy. This figure was drawn in-house, and reproduced with permission from Sato et al., "Sequential i-GONAD: An Improved In Vivo Technique for CRISPR/Cas9-Based Genetic Manipulations in Mice"; published by MDPI, 2020.

mouse production, which is time-consuming and labor intensive [130]. To bypass this tedious process, attempts to produce mice carrying *loxP*-floxed GOI (which are generally called "floxed mice") have been made using genome editing technology. In the initial experiments, simultaneous injection of Cas9, two pairs of gRNAs, and two ssODNs containing *lox* sequences into mouse zygotes generates mice containing floxed alleles [61, 131–134]. This approach can generate floxed mice without using ES cells, since it does not require the construction of a KI vector, and production of floxed mice is finished in a short period of time (e.g., in a month). However, according to Horii et al. [130], the simultaneous introduction of two mutated *lox* sites (to which Cre recombinase bind) at a target locus is difficult as it often causes LD of a sequence. To increase the possibility that the two *lox* sites are knocked-in, they used the sequential introduction approach to perform KI of mutated *lox* sites into the introns interposing exon 3 of methyl-CpG binding protein 2 gene (*Mecp2*) through *in vitro* EP. When the resulting embryos (blastocysts) were subjected to molecular analysis, 21% (33/155 tested) of the embryos had two floxed sites in the target *Mecp2* locus. Furthermore, the efficiency of generating LD was 36% (56/155 tested). Sato et al. [97] employed all the genome-editing components (gRNAs and ssODNs containing mutated *lox* sites as donor DNA) described by Horii et al. [130] for *si*-GONAD. Unfortunately, the generation of morulae with KI alleles (in which floxed sites had been knocked-in in both sides of the introns interposing exon 3 of *Mecp2*) failed; only a morula with one floxed site in the 5′ site of *Mecp2* was successfully generated.

Recently, Shang et al. [101] developed a new approach by integrating a unique design of asymmetric *loxP*-ssODN to create mouse conditional KO alleles in one step using the *i*-GONAD method. They injected a cocktail containing Cas9 protein, two gRNAs targeting the intron 2 and 3′ region of Fos-like antigen 1 gene (*Fosl1*), and two short ssODNs as HDR donors for *loxP* insertions. Each ssODN is 161 nucleotide (nt) long, composed of 91 nt of the 5′ homology arm from the PAM-proximal side, 34 nt of *loxP* sequence, and 36 nt of the 3′ homology arm from the PAM-distal side. Molecular biological analysis of the resulting pups demonstrated that out of 20 F_0 mice obtained one mouse had the simultaneous 5′- and 3′-*loxP* insertions and 6 had either 5′- or 3′-*loxP* integrations. Similar experiments were also conducted to obtain floxed mice for genes coding for pleomorphic adenoma gene-like 1 (*Plagl1*), *Ak040954*, cardiotrophin-like cytokine factor 1 (*Clcf1*), and *Gm4438*. The overall targeting efficiency of producing floxed alleles by *i*-GONAD was 10% (8/76 tested).

7.3 RFD is a useful reagent to master GONAD/*i*-GONAD in 2 days

RFD has been recognized as a useful reagent to judge the success of gene delivery to early mammalian embryos after *in vitro* EP [67]. It has also proved useful for judging the success of GONAD/*i*-GONAD [53, 86, 90, 91, 97, 119]. For example, when we performed *i*-GONAD using a solution containing RFD and TB (or Fast Green FC) at Day 0.7 of pregnancy (corresponding to late zygote stage; 16:00 PM), the success of the approach can be judged by examining the RFD-derived red fluorescence in the isolated 2-cell embryos under fluorescence microscopic observation one day after *i*-GONAD. The presence of fluorescent embryos means successful *i*-GONAD, while the absence of fluorescent embryos indicates failure of *i*-GONAD. This short-term experience is especially beneficial for the beginners who want to master the technique, and is thus called "2-day protocol for mastering GONAD" [117]. Notably, Sato et al. [97] showed that FITC-labeled fluorescent dextran is also effective for reporting the success of gene delivery in mouse early embryos.

7.4 Chromosomal engineering using *i*-GONAD

As shown in Section 7–1, CRISPR/Cas9-mediated genome editing using 2 types of gRNAs results in frequent generation of LD in the sequence flanked by the two sites recognized by these gRNAs. Iwata et al. [93] applied this technique to introduce chromosomal inversions of several megabases (Mb) in mice. When mouse zygotes were subjected to *in vitro* EP, a 7.67 Mb inversion was successfully introduced, which is longer than any previously reported inversion produced using MI-based methods.

They confirmed that a similar event can be induced using *i*-GONAD. These findings suggest that CRISPR/Cas9 system *via in vitro* and *in vivo* EP is useful for examining genetic diseases with large-scale chromosomal rearrangements.

Notably, the same group [107] recently demonstrated that *i*-GONAD can be useful to maintain lethal mutant mice using an inversion balancer identified from the C3H/HeJJcl strain. As a proof-of-principle, they created the Tp53rk binding protein gene (*Tprkb*) KO strain with an embryonic lethal mutation through *i*-GONAD in the presence of a non-targeted B6.C3H-In(6)1 J inversion. Iwata et al. [107] demonstrated that the edited lethal genes were stably maintained in heterozygotes, as recombination is strongly suppressed within this inversion interval. This strategy may facilitate further analysis of lethal mutants.

7.5 Viral transduction using *i*-GONAD

Virus-based gene delivery approaches have been widely used in the biomedical sciences, especially for gene therapeutic purposes (reviewed by Sung and Kim [135]). The viral vectors widely used are RV, AV, LV, and adeno-associated viral vector (AAV). Each of these vectors have specific properties. For example, RV and LV can infect both dividing and senescent cells and enable chromosomal integration. AV can infect mainly dividing cells efficiently, but cannot integrate into host chromosomes. These viral vectors (but not AAV) can infect ZP-removed early embryos, but not ZP-enclosed (or intact) embryos [13]. Notably, the simple incubation of ZP-enclosed embryos with recombinant AAV (rAAV)-containing medium was recently shown to lead to the transduction of those embryos [13, 92]. Notably, there are over 10 different serotypes of AAVs, each of which exhibits different infectious ability depending on the type of cells [136]. Mizuno et al. [13] examined which serotype of AAV could effectively transduce ZP-intact mouse 2-cell embryos. The embryos were co-incubated for 16 h with several types of rAAVs carrying an eGFP expression unit and then transferred to normal medium; the morulae developing after co-incubation with AAV serotype 6 (which is hereinafter called rAAV-6) exhibited strong fluorescence [13]. The next vector showing relatively strong infectivity was rAAV-1. A similar observation was also made by Yoon et al. [92]. Importantly, rAAV-6 can transduce rat and bovine embryos [13], suggesting the multi-species infectivity of this vector.

However, genome editing could not be induced in early embryos through transduction with rAAV-6. The rAAV carrying Cas9 gene (~9 kb in size) and that carrying gRNA could not be co-delivered, because rAAV was unable to incorporate over 4.5-kb of an insert [137]. To perform successful CRISPR-based KI in mice, Mizuno et al. [13] employed a two-step gene delivery approach. Zygotes were first subjected to *in vitro* EP in the presence of RNP and then transduced with rAAV-6 carrying a 1.8-kb GFP expression cassette flanked by two 100-bp *Rosa26* homology arms. Molecular biological analysis of the newborn pups demonstrated that the KI efficiency in the *Rosa26* locus was 6%. Yoon et al. [92] performed intraoviductal injection of a solution containing rAAV6 Cas9 (carrying *spCas9* gene derived from *Streptococcus pyogenes*) and rAAV6-gTyr (carrying gRNA expression unit targeted *Tyr*) into the pregnant female mice at Day 0.5 of pregnancy, similar to GONAD/*i*-GONAD. Molecular biological analysis of the newborn pups demonstrated that the indel efficiency was 6%. All mutated founder (F_0) mice generated albino offspring, indicating germ-line transmission; this suggested that AAV is a powerful tool for inducing genome editing in the ZP-enclosed early embryos *in vivo*. According to Sato et al. [138], this *in vivo* approach is referred to as "AAV-based GONAD."

As was mentioned earlier, AAV-based GONAD appears to be more convenient than GONAD/*i*-GONAD, since the former does not require any of the apparatus required for EP. Unfortunately, it still requires more detailed information concerning (1) which rAAV serotype is effective for *in vivo* transduction towards early mouse embryos, (2) the stage allowing maximal expression of GOI (included in rAAV) after infection at late zygote stage, and (3) whether the oviductal epithelial cells are infected through AAV-based GONAD. Sato et al. [139] first examined the above possibility using 4 types (1, 2, 5, and 6) of rAAVs carrying a unit for expression of eGFP (as GOI). When AAV-based GONAD was performed at Day 0.7 of pregnancy, and 2 days later the morulae were isolated to inspect eGFP fluorescence, rAAV-6 gave strongest fluorescence, though the fluorescence intensity varied among embryos. The fluorescence intensity provided by rAAV-1 was the next highest, but transduction with the other remaining serotypes (2 and 5) resulted in negative or faint fluorescence in the embryos. These results are consistent with the *in vitro* data from Mizuno et al. [13] and Yoon et al. [92]. A similar mode of transduction was also seen in the oviductal epithelial cells, suggesting the use of rAAV-6 (or possibly rAAV-1) for genome manipulation of those cells.

AAV-based GONAD using rAAV-6 was performed at Day 0.7 of pregnancy, and one day later, 2-cell embryos were isolated and cultured until the late blastocyst stage to monitor the eGFP fluorescence expression. Under the fluorescence microscope, fluorescence was first discernible at the 2-cell stage, attained at a maximal level at the morula stage, and declined towards late blastocyst stage [139]. These results suggest that one-day infection with rAAV-6 is enough to transduce ZP-enclosed zygotes floating in the oviductal lumen. Furthermore, the GOI expression was transient, peaking at the morula stage. These findings suggest a possibility that early mouse embryos from zygote to morula stages can be effectively transduced in a sequential manner, like *si*-GONAD.

7.6 Effect of HA treatment on the efficiency of *i*-GONAD-mediated genome editing

As was mentioned in Section 4, *i*-GONAD at Day 0.4 of pregnancy (corresponding to early zygote) often failed to obtain genome-edited embryos/fetuses. This appears to be solely due to the presence of cumulus cells that tightly surround a zygote. One idea to overcome this problem may be that early zygotes are pretreated with HA, an enzyme capable of dispersing cumulus cells from a zygote [1], prior to *i*-GONAD. Kaneko and Tanaka [100] examined the possibility by injecting 1 μL of 0.1% HA into the ampulla of a female (ICR) at Day 0.4 of pregnancy (10:00–11:00) using a thin glass needle. As a control, the solvent (PBS) was similarly injected. Several minutes after the injection, a solution (1 μL) containing genome editing reagents (2 μg/μL Cas9 protein +60 mM dual gRNA (targeted fibroblast growth factor 10 gene (*Fgf10*) + 0.08% TB) was intraoviductally introduced and subsequently the entire oviducts were subjected to *in vivo* EP using tweezer-type electrodes. After that, the developing fetal offspring were isolated for examining the presence of possible genome editing in those samples. Consequently, the samples isolated from HA-treated group exhibited 2.5-fold higher genome editing (indels) efficiency than those isolated from the control group (68% *vs.* 27%). The *i*-GONAD on Day 0.7 of pregnancy (16:00–17:00; in which case no HA is used) yielded genome edited pups with an efficiency of 54%. These findings indicate that HA-mediated removal of cumulus cells at Day 0.4 of pregnancy is effective when *in situ* genome editing towards early zygotes are intended. According to Kaneko and Tanaka [100], the operation time for introducing genome editing reagents into embryos in the oviducts can be adjusted by treatment with HA before EP. This improved protocol can also be used for efficient production of genome-edited mice and rats.

7.7 Strain-difference can affect the efficiency of *i*-GONAD-mediated genome editing

According to Ohtsuka et al. [86], successful *i*-GONAD relies on the mouse strain used. For example, it worked successful under relatively stringent electrical conditions (40 V/100–200 Ω/~300 mA) when random-bred mice (such as MCH(ICR) and B6C3F1, a hybrid between C3H/He and C57BL/6), but not C57BL/6 strain, were used. Under less stringent conditions (40 V/350–400 Ω/~100 mA), *i*-GONAD was successful in the inbred C57BL/6 strain [86, 98, 117]. These findings suggest the importance of selecting the appropriate EP conditions, particularly when different mouse strains are used for *i*-GONAD experiment.

This is also true for *i*-GONAD using rats. For example, when a current of >500 mA was employed using the NEPA21 electroporator, albino SD and albino LEW rats were successfully genome-edited; however, no offspring were derived from pigmented BN rats (fetuses/newborns) [91]. In contrast, *i*-GONAD was performed under a current of 100–300 mA using the NEPA21 electroporator, leading to the production of genome-edited BN rats at efficiencies of 75–100% [99]. Similar success in producing GM BN rats was achieved with efficiencies of 24–55% when another electroporator CUY21EDIT II (BEX Co., Ltd., Tokyo, Japan) was employed under a current of 150–200 mA [99].

Notably, the most widely used electroporators (as exemplified by NEPA21) employ a constant voltage. Also, other electroporators (as exemplified by GEB15 (BEX Co., Ltd.)) employ a constant current. Kobayashi et al. [98] explored the conditions allowing the generation of a 100 mA current in C57BL/6 mice using two electroporators, NEPA21 and GEB15. As a result, *i*-GONAD performed under conditions of average resistance of 367 Ω and average voltage of 116 mA resulted in the production of genome-edited fetuses with efficiency of 39%.

7.8 Attempt to increase the efficiency of KI using *i*-GONAD

In our previous study using *i*-GONAD to produce GM rats and KO/KI rats, the success rate of producing KI rats was lower than that of KI mice (5% *vs.* 60%, respectively) when ssODNs were used as KI donors [91, 117]. To improve the efficiency of *i*-GONAD in rats, Aoshima et al. [112] examined the effects of commercially available KI-enhancing drugs (including SCR7, L755,507, RAD51-stimulatory compound 1 (RS-1) and Alt-R® HDR Enhancer (HDR enhancer)), some of which have been known to increase KI efficiency in culture cells and early embryos [140–143]. For example, *i*-GONAD was applied to SD female rats (albino) using a solution containing RNP complex (consisting of Cas9 protein and gRNA targeted *Tyr* locus), ssODN (used as a KI donor oligodeoxynucleotide), and various amounts (5 or 15 μM) of L755,507 on Day 0.7 of pregnancy. Inspection of mid-gestational fetuses revealed that 12% of fetuses obtained showed pigmented eyes when 5 μM L755,507 was used for *i*-GONAD, suggesting successful KI [112]. In addition to L755,507, some drugs (e.g., SCR7 and HDR enhancer) were found to be effective in *i*-GONAD in rats, but their effects were limited.

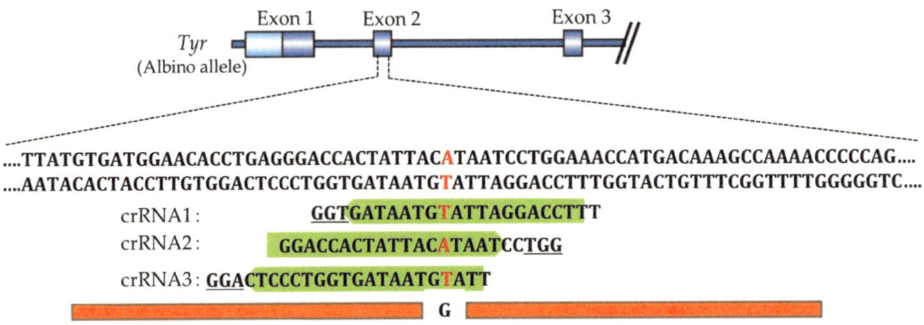

Figure 5.
Schematic of knock-in (KI) experiment in rats towards the mutated tyrosinase gene (Tyr) locus performed by Aoshima et al. [112]. The target sequence (exon 2 of Tyr) recognized by crRNA1, 2, and 3 is shown in green. PAM sequences are underlined. Single-stranded oligodeoxynucleotide (ssODN) (containing wild-type nucleotide "G" that corresponds to mutated nucleotide "A") is shown in orange below the target sequence. The nucleotide "A/T" marked in red is the mutation causing the albino phenotype. This figure was drawn in-house, and reproduced with permission from Aoshima et al., "Modification of improved-genome editing via oviductal nucleic acids delivery (i-GONAD)-mediated knock-in in rats"; published by BioMed Central Ltd, 2021.

In a study by Aoshima et al. [112], three gRNAs (called crRNA1, crRNA2, and crRNA3) were used. As shown in **Figure 5**, these gRNAs recognize different portions of the target locus, but also overlap each other in the target locus. Surprisingly, the KI efficiency in rat fetuses generated after *i*-GONAD with crRNA2 and ssODN was significantly higher (24%) than crRNA1 (5%) or crRNA3 (0%). The KI efficiency of *i*-GONAD with triple gRNAs was 11%. These findings demonstrated that the choice of gRNA is important for determining KI efficiency.

7.9 Regulated timing of *i*-GONAD by administration of gonadotrophins

The *i*-GONAD experiment using C57BL/6 strain is always associated with the difficulty in consistently obtaining pregnant females, because estrous females are not always available. The administration of gonadotrophins has been frequently used for inducing superovulation in many mouse strains to obtain a number of early embryos [1]. This approach has an additional advantage in that it is capable of synchronizing the estrous cycle of females; thus, the estrous cycle need not be examined through smear testing or through visual inspection of the vagina.

Administration of higher dose (in this case, more than 5 international units (IU)) of gonadotrophins can induce superovulation, but often causes failure to deliver pups [144–146]. Administering low doses (less than 5 IU) of gonadotrophins facilitates ovulation of natural number of oocytes and successful delivery of pups [147, 148]. Notably, Sato et al. [53] reported that intraperitoneal (IP) administration of low-dose (2–0.5 IU) serum gonadotrophin (PMSG) from a pregnant mare on 11:00, followed by 5 IU of human chorionic gonadotrophin (hCG) 48 h later, is effective for inducing natural ovulation before *i*-GONAD. In case of administration of 5 IU PMSG, females having vaginal plug failed to deliver their pups. When females were inspected later, some had dead fetuses in their uterus. When 2 or 0.5 IU PMSG was administered, all females successfully delivered viable pups (average: 8 in each group). These findings suggest that *i*-GONAD can be performed on 11:00 at Day 0.7 of pregnancy when females were induced to ovulate by administering a low dose of PMSG. Indeed, Sato et al. [97] demonstrated that *i*-GONAD on 11:00 at Day 0.7 of pregnancy leads to generation of genome-edited morulae.

Unfortunately, the regime shown by Sato et al. [53] has only been used successfully on B6C3F1 hybrid mice. Kobayashi et al. [98] examined whether the administration of a single IP injection of low-dose PMSG (1.2 IU/10 g) is effective for synchronizing the estrous cycle in C57BL/6 females. Consequently, approximately 51% of C57BL/6 females had plugs upon mating with males 2 days after PMSG administration, which contrasts with the case (~26%) when C57BL/6 females were subjected to natural mating. Furthermore, 44% hormone-injected and plugged females delivered pups with an average litter size of six, which was comparable to the rate obtained from females that were not injected with hormones. These findings indicate that a single IP injection of low-dose PMSG increases the rate of acquiring plugged females before mating. This is particularly beneficial for *i*-GONAD which always requires desired number of plugged females obtained through a scheduled mating.

7.10 Combinational use of *i*-GONAD with anti-inhibin monoclonal antibodies (AIMAs) treatment to increase the number of GM mice

Many attempts have been made to increase litter sizes (which is determined by the number of oocytes naturally ovulated) using conventional superovulation regimens (e.g., using PMSG/hCG), but had limited success because of unexpected decreases in the numbers of embryos surviving to term, as mentioned in the Section 7–9. Hasegawa et al. [115] attempted to overcome this problem using rat-derived AIMAs. They administrated progesterone (P4) once a day for 2 days (days 1 and 2) to synchronize the estrous cycle of female C57BL/6 mice, and AIMAs were injected into the same animals at Day 4 followed by mating with male C57BL/6 mice. When *i*-GONAD targeting *Tyr* was applied to the AIMA-treated C57BL/6 female mice on the day of vaginal plug formation during Days 6–8, a 1.5-fold increase in litter size was observed (7.3 *vs.* 4.8 for the untreated control). Notably, genome editing efficiency did not differ between these two groups. Therefore, AIMA treatment can reduce the number of females used for the *i*-GONAD experiment, which will fulfill the 3R principles of animal experimentation (i.e., Reduction, Replacement, and Refinement), and can be applied to other mouse strains and animals.

7.11 GONAD/*i*-GONAD as a useful tool to check *in vivo* gene correction event

As was shown previously, GONAD/*i*-GONAD enable gene delivery to early embryos present within the oviductal lumen and to the epithelial cells facing the lumen [37, 53, 139]. The success of gene delivery to the oviductal epithelial cells can be easily judged through direct observation of fluorescence under a fluorescence microscope [37, 53, 139]. Therefore, *in vivo* gene delivery approach targeting oviductal epithelial cells are excellent for testing the function of the GOI.

Miura et al. [109] attempted to examine whether *in vivo* CRISPR/Cas9-mediated gene correction is possible using GONAD/*i*-GONAD technologies. They first generated Δ*eGFP* KO mouse strain through MI of a solution containing *Cas9* mRNA and gRNA (targeted *eGFP* cDNA) into zygotes from Tg mice carrying *eGFP* cDNA [85]. The resulting Δ*eGFP* KO mice failed to exhibit systemic eGFP expression, due to frame-shift mutations in the coding region of *eGFP* cDNA that was chromosomally integrated in their genome. Next, they injected a solution containing sgRNA targeted to the mutation site, Cas9 protein, ssODN (as donor DNA sequence; in some cases), and Fast Green FCF into the oviductal lumen of female Δ*eGFP* KO mouse. Subsequently, the entire oviducts were electroporated, similar to that performed in

GONAD/*i*-GONAD. Three to 13 days later, the eGFP fluorescence was inspected in the oviducts dissected from the treated females. Consequently, fluorescence was detected in a portion of an oviduct, suggesting gene editing at the mutated site in the *eGFP* cDNA through HDR-mediated KI of ssODN or NHEJ-mediated indels. Molecular biological analysis of the oviduct confirmed the above events. Notably, in this system, editing of mutated site can be easily monitored by visually inspecting the gene-edited oviducts under UV illumination.

7.12 GONAD/*i*-GONAD as a useful tool to generate mouse models for ovarian cancer

As was mentioned previously, GONAD/*i*-GONAD could transfect both pre-implantation embryos and the oviductal epithelium facing the oviductal lumen. Ovarian cancer is the most lethal gynecologic cancer to date. High-grade serous ovarian carcinoma (HGSOC) is the most common type of ovarian cancer and has the lowest rate of survival. Teng et al. [149] recreated the mutations found in ovarian cancer to generate somatic ovarian cancer mouse models, using an *in vivo* oviductal EP method similar to GONAD/*i*-GONAD. Using the CRISPR/Cas9 genome editing approach, they mutated the tumor suppressor genes (transformation related protein 53 (*Trp53*), breast cancer susceptibility gene I (*Brca1*), neurofibromin 1 (*Nf1*), and phosphatase and tensin homolog (*Pten*)) to study how these genes contribute to tumor development. When mutations were introduced in three of the four genes, namely *Trp53*, *Brca1*, and either *Nf1* or *Pten*, the sites transfected with the genome editing reagents displayed effects that were similar to human HGSOC and changes in chromosome number. Teng et al. [149] concluded that the *in vivo* oviductal EP method is highly useful for generating mouse models to advance the understanding and treatment of ovarian cancer.

8. Challenges, limitations, concerns and future perspective

8.1 Challenges

GONAD/*i*-GONAD is performed by injecting a small amount (1–2 μL) of a solution containing genome editing components into the specific site of an oviduct (called ampulla) of a pregnant female at Day 0.7 to 1.4 of pregnancy, using oral breath-controlled glass micropipette under a dissecting microscope. The genome editing components introduced within an oviductal lumen exist around ZP-enclosed early embryos (zygotes to 2-cell embryos), but are never incorporated into those embryos in an intact state. However, *in vivo* EP enables delivery of substances (present outside the embryos) into the internal portion of an embryo, leading to generation of genome-edited embryos. Another method to induce genome editing in early embryos *in situ* is viral transduction using rAAV. Unfortunately, EP requires expensive apparatus such as an electroporator, and rAAV transduction requires labor-intensive and time-consuming preparation of viral particles. In this context, the use of ZP-penetrating agents (e.g., MWNT and VisFect) would be ideal, because it does not require an electroporator or viral preparation. To date, there is no report on successful genome editing in early embryos using these ZP-penetrating agents. We expect that these agents could be useful for facilitating NA delivery to embryos and subsequent induction of genome editing.

Kaneko and Tanaka [100] demonstrated that pretreatment of early zygotes (tightly surrounded by cumulus cells) with HA led to increased efficiency of genome editing in those embryos after *i*-GONAD. This is based on the concept that cumulus cells surrounding a zygote hamper rapid transfer of genome editing reagents to zygotes [86]. In the previous approach using GONAD/*i*-GONAD, *in vivo* EP is applied immediately after intraoviductal injection of NAs [84, 86, 116]. In this case, it is highly likely that the reagents injected might not have been fully infiltrated between the intercellular space connecting cumulus cells. Waiting for several minutes after intraoviductal injection may permit sufficient infiltration of the reagents before *in vivo* EP, leading to increased efficiency of genome editing. This line of experiment has now been undertaken by Takabayashi and his colleagues.

8.2 Limitations

GONAD/*i*-GONAD can be applied to larger animals, as exemplified by its use in pigs; the demand for GM pigs has been rising due to the needs of the biomedical and agricultural fields [150, 151]. The current strategy for creating GM pigs is based on "*ex vivo* handling of embryos," where MI or EP is carried *in vitro* towards zygotes collected from individuals or produced through IVF, following which the treated embryos have to be subjected to ET to recipient females [138]. Similar to the case of MI or *in vitro* EP-mediated production of GM mice and rats, creation of GM pigs is highly costive, time-consuming, and labor-intensive. Additionally, it requires recipient female pigs, which are also expensive. In this context, GONAD/*i*-GONAD should be theoretically performed in more convenient and inexpensive manner, since it does not require "*ex vivo* handling of embryos." Notably, the porcine oviduct is generally ca. 100 mm in length and their form is linear, unlike the spiral form in rodents (mice and rats). Furthermore, it does not have an enlarged site called "ampulla" where rodent zygotes always stay. As a result, porcine zygotes will exist in a broad area throughout the oviduct. To perform GONAD/*i*-GONAD in pigs, researchers must inject a large amount of fluid (probably over 1 mL) and electroporate towards several sites whenever tweezer-type electrodes are used. It remains uncertain whether this is feasible. Therefore, GONAD/*i*-GONAD in pigs remains a challenge.

8.3 Concerns

As was mentioned in Section 1, in the initial step of development of GONAD/*i*-GONAD, gene delivery to early murine embryos using plasmid DNA were successful when the T-820 electroporator from BTX Co. was used [28]. However, it was impossible when the other electroporators such as NEPA21 and CUY21EDIT II were employed (Sato, Ohtsuka, unpublished). Notably, Hakim et al. [54] recently checked several *in vitro* EP parameters to seek optimal conditions enabling gene (plasmid DNA) delivery into mouse follicles, oocytes, and early embryos. When they were electroporated in the 1-mm gap cuvettes using Gene Pulser Mxcell System (Bio-rad Laboratories, Hercules, CA, USA), EP under 3 pulses of 30 V of 1 ms each at an interval of 10 s was ideal, with no need to weaken or loosen the ZP layer. This suggests that exploration of optimal EP condition using the above apparatuses (NEPA21 and CUY21EDIT II) may enable the transfection of ZP-intact embryos with plasmid DNA. If it is realized, transgenesis *via* introduction of plasmid-based transposons (e.g., *piggy*Bac (PB) transposon + PB transposase mRNA) may be possible.

8.4 Future perspective

Preimplantation embryos at zygote to morula stages, all of which exist within the oviductal lumen of a pregnant female, can be a target for gene delivery through GONAD/*i*-GONAD. As cleavage embryos (at 2-cell to early 8-cell stages) are comprised of blastomeres, each of which is facing the outside environment, genome editing at these stages may frequently result in the generation of mosaic offspring containing edited and unedited cells. This mosaic nature is tedious for the researchers who want to produce GM animals with high efficiency, but in turn beneficial for investigating the properties of embryonic lethal genes, because mosaic fetuses or pups produced through MI of genome editing components into one blastomere of two-cell embryos should be viable and carry heritable lethal mutations [152]. On the other hand, in the case of compacted 8-cell embryos and morulae, only blastomeres facing the external environment (but not the inner cells present inside an embryo) can be susceptible to genome editing. The outer blastomeres of morulae are thought to contribute to the formation of a trophectodermal cell, which is a cell involving implantation and placenta formation. Therefore, the GONAD/*i*-GONAD-mediated genome manipulation at these stages may be a novel tool to explore the molecular mechanisms underlying implantation and placenta formation.

9. Conclusion

Seven years have passed since the first report [84] on the development of GONAD using mice. During this period, successful genome editing was reported in other animals including rats and hamsters. The genes targeted by GONAD/*i*-GONAD technologies were *eGFP* cDNA (chromosomally integrated in the *eGFP* Tg mice) [84, 109] and endogenous genes such as *Acrosin*, *Agouti*, *Ak040954*, *ATF5*, *Axdnd 1*, *Cdkn1a*, *Cdkn2a*, *Clcf1*, *COL4A3*, *COL4A4*, *COL4A5*, *COL7A1*, *Dll1*, *eEF2*, *Fgf10*, *Fosl1*, *Foxe3*, *Gbx2*, *GGTA1*, *Gm30413*, *Gm44386*, *Hprt*, *Kit*, *Klrc2*, *Mecp2*, *Mov10l1*, *Nrsn2*, *Pafah1b1*, *Pax6*, *Pitx3*, *Plagl1*, *Serpina3n*, *Tis21*, *Tprkb*, and *Tyr* (**Table 1**). All these genes were disrupted by indel mutations (KO) or modified through HDR-mediated KI using the CRISPR/Cas9 system. As for the components used for CRISPR/Cas9-mediated genome editing, *Cas9* mRNA was initially employed, but later Cas9 protein was mainly used. For KI experiments, ssODN or *in vitro* synthesized long ssDNA was used as donor DNA. The dye used for monitoring successful injection process was also changed; while TB was initially employed, Fast Green FCF was used for later experiments. According to Ohtsuka et al. [86], TB (but not Fast Green FCF) often generates visible precipitates, when RNP is mixed with the dye. GONAD/*i*-GONAD technologies are very simple systems that only require the intraoviductal injection of NA-containing solution and subsequent *in vivo* EP. EP is now recognized as a powerful tool enabling efficient gene delivery, but often causes deleterious effects on cell/embryo survival, leading to reduction in litter size, as suggested by Kaneko and Tanaka [100]. Ohtsuka et al. [86] first demonstrated that the success of *in vivo* EP depends on the mouse strains used. This was also true for generating GM rats using *i*-GONAD [91]. Therefore, exploration of optimal EP conditions is required before GONAD/*i*-GONAD experiments start using new strains.

GONAD/*i*-GONAD requires an expensive electroporator. Notably, the intraoviductal injection of rAAV into a pregnant female mouse was also useful for the *in situ* transduction of zygotes [92, 139]. Furthermore, it has already been shown that

AAV-mediated genome editing at zygotes through AAV-based GONAD is powerful for the convenient acquisition of GM animals [92]. In this AAV-based GONAD system, the electroporator need not be used, although it is always associated with laborious and time-consuming tasks such as viral vector preparation. For developing systems that are simpler than the present GONAD/*i*-GONAD, the use of ZP-penetrating reagents will be highly desirable.

Recently, new genome editing systems known as prime editing [153–156] and base editing [157–159] were reported. These systems enable precise gene correction at a single nucleotide level. To date, these reagents have not been used for GONAD/*i*-GONAD-mediated production of GM animals. Future application of these new genome editing systems to GONAD/*i*-GONAD is highly expected.

Acknowledgements

We thank Kazusa Inada for her support for in-house drawing of the Figures, shown in **Figures 1–4**. This study was partly supported by a grant (no. 24580411 for Masahiro Sato; no. 21 K10165 for Emi Inada; no. 16H05049 for Shingo Nakamura; no. 22H03277 for Issei Saitoh; no. 16 K07087 for Shuji Takabayashi) from the Ministry of Education, Science, Sports, and Culture, Japan, and a fund for the Promotion of Joint International Research (Fostering Joint International Research) (no. 16KK0189 for Masato Ohtsuka) from JSPS, Japan.

Conflict of interest

The founding sponsors had no role in the design of the study, collection, analyses, or interpretation of data, writing of the manuscript, and decision to publish the results.

Notes

Masahiro Sato and Shuji Takabayashi designed the study and drafted the manuscript; Masato Ohtsuka, Emi Inada, Shingo Nakamura and Issei Saitoh critically revised the manuscript.

Author details

Masahiro Sato[1*], Masato Ohtsuka[2,3,4], Emi Inada[5], Shingo Nakamura[6], Issei Saitoh[7] and Shuji Takabayashi[8]

1 Department of Genome Medicine, National Center for Child Health and Development, Tokyo, Japan

2 Department of Molecular Life Science, Division of Basic Medical Science and Molecular Medicine, Tokai University School of Medicine, Isehara, Kanagawa, Japan

3 Center for Matrix Biology and Medicine, Graduate School of Medicine, Tokai University, Isehara, Kanagawa, Japan

4 The Institute of Medical Sciences, Tokai University, Isehara, Kanagawa, Japan

5 Department of Pediatric Dentistry, Graduate School of Medical and Dental Sciences, Kagoshima University, Kagoshima, Japan

6 Division of Biomedical Engineering, National Defense Medical College Research Institute, Saitama, Japan

7 Department of Pediatric Dentistry, Asahi University School of Dentistry, Mizuho, Japan

8 Laboratory Animal Facilities and Services, Preeminent Medical Photonics Education and Research Center, Hamamatsu University School of Medicine, Hamamatsu, Shizuoka, Japan

*Address all correspondence to: sato-masa@ncchd.go.jp

References

[1] Hogan B, Beddington R, Constantini F, Lacy E, editors. Manipulating the Mouse Embryo. 2nd ed. Cold Spring Harbor, NY (USA): Cold Spring Harbor Laboratory Press; 1994

[2] Jaenisch R, Mintz B. Simian virus 40 DNA sequences in DNA of healthy adult mice derived from preimplantation blastocysts injected with viral DNA. Proceedings of the National Academy of Sciences of the United States of America. 1974;**71**:1250-1254. DOI: 10.1073/pnas.71.4.1250

[3] Gordon JW, Scangos GA, Plotkin DJ, Barbarosa JA, Ruddle FH. Genetic transformation of mouse embryos by microinjection of purified DNA. Proceedings of the National Academy of Sciences of the United States of America. 1980;**77**:7380-7384. DOI: 10.1073/pnas.77.12.7380

[4] Hammer RE, Pursel VG, Rexroad CE Jr, Wall RJ, Bolt DJ, et al. Production of transgenic rabbits, sheep and pigs by microinjection. Nature. 1985;**315**:680-683. DOI: 10.1038/315680a0

[5] Perry AC, Wakayama T, Kishikawa H, Kasai T, Okabe M, et al. Mammalian transgenesis by intracytoplasmic sperm injection. Science. 1999;**284**:1180-1183. DOI: 10.1126/science.284.5417.1180

[6] Perry AC, Rothman A, de las Heras JI, Feinstein P, Mombaerts P, et al. Efficient metaphase II transgenesis with different transgene archetypes. Nature Biotechnology. 2001;**19**:1071-1073. DOI: 10.1038/nbt1101-1071

[7] Moreira PN, Giraldo P, Cozar P, Pozueta J, Jimenez A, et al. Efficient generation of transgenic mice with intact yeast artificial chromosomes by intracytoplasmic sperm injection. Biology of Reproduction. 2004;**71**:1943-1947. DOI: 10.1095/biolreprod.104.032904

[8] Osada T, Toyoda A, Moisyadi S, Akutsu H, Hattori M, et al. Production of inbred and hybrid transgenic mice carrying large (> 200 kb) foreign DNA fragments by intracytoplasmic sperm injection. Molecular Reproduction and Development. 2005;**72**:329-335. DOI: 10.1002/mrd.20319

[9] Wassarman PM. Zona pellucida glycoproteins. The Annual Review of Biochemistry. 1988;**57**:415-442. DOI: 10.1146/annurev.bi.57.070188.002215

[10] Clift D, Schuh M. Re-starting life: Fertilization and the transition from meiosis to mitosis. Nature Reviews Molecular Cell Biology. 2013;**14**:549-562. DOI: 10.1038/nrm3643

[11] Tsukui T, Kanegae Y, Saito I, Toyoda Y. Transgenesis by adenovirus-mediated gene transfer into mouse zona-free eggs. Nature Biotechnology. 1996;**14**:982-985. DOI: 10.1038/nbt0896-982

[12] Gordon JW. High toxicity, low receptor density, and low integration frequency severely impede the use of adenovirus vectors for production of transgenic mice. Biology of Reproduction. 2002;**67**:1172-1179. DOI: 10.1095/biolreprod67.4.1172

[13] Mizuno N, Mizutani E, Sato H, Kasai M, Ogawa A, et al. Intra-embryo gene cassette knockin by CRISPR/Cas9-mediated genome editing with adeno-associated viral vector. iScience. 2018;**9**:286-297. DOI: 10.1016/j.isci.2018.10.030

[14] Kubisch HM, Larson MA, Lichen PA, Wilson JM, Roberts KM. Adenovirus-mediated gene transfer by perivitellin microinjection of mouse, rat, cow embryos. Biology of Reproduction. 1997;**56**:119-124. DOI: 10.1095/biolreprod56.1.119

[15] Lois C, Hong EJ, Pease S, Brown EJ, Baltimore D. Germline transmission and tissue-specific expression of transgenes delivered by lentiviral vectors. Science. 2002;**295**:868-872. DOI: 10.1126/science.1067081

[16] Martin NP, Myers P, Goulding E, Chen SH, Walker M, et al. En masse lentiviral gene delivery to mouse fertilized eggs via laser perforation of zona pellucida. Transgenic Research. 2018;**27**:39-49. DOI: 10.1007/s11248-017-0056-8

[17] Carballada R, Degefa T, Esponda P. Transfection of mouse eggs and embryos using DNA combined to cationic liposomes. Molecular Reproduction and Development. 2000;**56**:360-365. DOI: 10.1002/1098-2795(200007)56:3<360::AID-MRD5>3.0.CO;2-8

[18] Ikawa M, Tanaka N, Kao WW, Verma IM. Generation of transgenic mice using lentiviral vectors: A novel preclinical assessment of lentiviral vectors for gene therapy. Molecular Therapy. 2003;**8**:666-673. DOI: 10.1016/s1525-0016(03)00240-5

[19] Nijs M, Van Steirteghem AC. Assessment of different isolation procedures for blastomeres from two-cell mouse embryos. Human Reproduction. 1987;**2**:421-424. DOI: 10.1093/oxfordjournals.humrep.a136561

[20] Bronson RA, McLaren A. Transfer to the mouse oviduct of eggs with and without the zona pellucida. Journal of Reproduction and Fertility. 1970;**22**:129-137. DOI: 10.1530/jrf.0.0220129

[21] Modlinski JA. The role of the zona pellucida in the development of mouse eggs *in vivo*. Journal of Embryology and Experimental Morphology. 1970;**23**:539-547

[22] Carballada R, Relloso M, Esponda P. Generation of transgenic mice by transfection of pronuclear embryos using lipid-DNA complexes. Zygote. 2002;**10**:209-216. DOI: 10.1017/s0967199402002277

[23] Jaroszeski MJ, Gilbert R, Nicolau C, Heller R. *In vivo* gene delivery by electroporation. Advanced Drug Delivery Reviews. 1999;**35**:131-137. DOI: 10.1016/S0169-409X(98)00068-4

[24] Nemec LA, Skow LC, Goy JM, Kraemer DC. Introduction of DNA into murine embryos by electroporation. Theriogenology. 1989;**31**:233. DOI: 10.1016/0093-691X(89)90641-9

[25] Grabarek JB, Plusa B, Glover DM, Zernicka-Goetz M. Efficient delivery of dsRNA into zona-enclosed mouse oocytes and preimplantation embryos by electroporation. Genesis. 2002;**32**:269-276. DOI: 10.1002/ gene.10076

[26] Wang H, Ding T, Brown N, Yamamoto Y, Prince LS, et al. Zonula occludens-1 (ZO-1) is involved in morula to blastocyst transformation in the mouse. Developmental Biology. 2008;**318**:112-125. DOI: 10.1016/j.ydbio.2008.03.008

[27] Peng H, Wu Y, Zhang Y. Efficient delivery of DNA and morpholinos into mouse preimplantation embryos by electroporation. PLoS One. 2012;**7**:e43748. DOI: 10.1371/journal.pone.0043748

[28] Sato M, Akasaka E, Saitoh I, Ohtsuka M, Watanabe S. *In vivo* gene

transfer in mouse preimplantation embryos after intraoviductal injection of plasmid DNA and subsequent *in vivo* electroporation. Systems Biology in Reproductive Medicine. 2012;**58**:278-287. DOI: 10.3109/19396368.2012.688088

[29] Ivanova MM, Rosenkranz AA, Smirnova OA, Nikitin VA, Sobolev AS, et al. Receptor-mediated transport of foreign DNA into preimplantation mammalian embryos. Molecular Reproduction and Development. 1999;**54**:112-120. DOI: 10.1002/(SICI)1098-2795(199910)54:2<112::AID-MRD2>3.0.CO;2-U

[30] Heyner S, Rao LV, Jarett L, Smith RM. Preimplantation mouse embryos internalize maternal insulin via receptor-mediated endocytosis: Pattern of uptake and functional correlations. Developmental Biology. 1989;**134**:48-58. DOI: 10.1016/0012-1606(89)90077-8

[31] Joo JY, Lee J, Ko HY, Lee YS, Lim D-H, et al. Microinjection free delivery of miRNA inhibitor into zygotes. Scientific Reports. 2014;**4**:5417. DOI: 10.1038/srep05417

[32] Munk M, Ladeira LO, Carvalho BC, Camargo LSA, Nádia RB, et al. Efficient delivery of DNA into bovine preimplantation embryos by multiwall carbon nanotubes. Scientific Reports. 2016;**6**:33588. DOI: 10.1038/srep33588

[33] Jin Z, Li R, Zhou C, Shi L, Zhang X, et al. Efficient gene knockdown in mouse oocytes through peptide nanoparticle-mediated siRNA transfection. PLoS One. 2016;**11**:e0150462. DOI: 10.1371/journal.pone.0150462

[34] Relloso M, Esponda P. *In vivo* gene transfer to the mouse oviduct epithelium. Fertility and Sterility. 1998;**70**:366-368. DOI: 10.1016/s0015-0282(98)00144-7

[35] Relloso M, Esponda P. *In vivo* transfection of the female reproductive tract epithelium. Molecular Human Reproduction. 2000;**6**:1099-1105. DOI: 10.1093/molehr/6.12.1099

[36] Rios M, Venegas A, Croxatto HB. *In vivo* expression of beta-galactosidase by rat oviduct exposed to naked DNA or messenger RNA. Biological Research. 2002;**35**:333-338. DOI: 10.4067/s0716-97602002000300007

[37] Sato M. Intraoviductal introduction of plasmid DNA and subsequent electroporation for efficient in vivo gene transfer to murine oviductal epithelium. Molecular Reproduction and Development. 2005;**71**:321-330. DOI: 10.1002/mrd.20295

[38] Osumi N, Inoue T. Gene transfer into cultured mammalian embryos by electroporation. Methods. 2001;**24**:35-42. DOI: 10.1006/meth.2001.1154

[39] Nakamura H, Isaka Y, Tsujie M, Akagi Y, Sudo T, et al. Electroporation-mediated PDGF receptor-IgG chimera gene transfer ameliorates experimental glomerulonephritis. Kidney International. 2001;**59**:2134-2145. DOI: 10.1046/j.1523-1755.2001.00728.x

[40] Tsujie M, Isaka Y, Nakamura H, Imai E, Hori M. Electroporation-mediated gene transfer that targets glomeruli. Journal of the American Society of Nephrology. 2001;**12**:949-954. DOI: 10.1681/ASN.V125949

[41] Heller R, Jaroszeski M, Atkin A, Moradpour D, Gilbert R, et al. *In vivo* gene electroinjection and expression in rat liver. FEBS Letters. 1996;**389**:225-228. DOI: 10.1016/0014-5793(96)00590-x

[42] Nishi T, Yoshizato K, Yamashiro S, Takeshima H, Sato K, et al. High-efficiency *in vivo* gene transfer

using intraarterial plasmid DNA injection following *in vivo* electroporation. Cancer Research. 1996;**6**:1050-1055

[43] Reiss M, Jastreboff MM, Bertino JR, Narayanan R. DNA-mediated gene transfer into epidermal cells using electroporation. Biochemical and Biophysical Research Communications. 1986;**137**:244-249. DOI: 10.1016/0006-291x(86)91202-7

[44] Titomirov AV, Sukharev S, Kistanova E. *In vivo* electroporation and stable transformation of skin cells of newborn mice by plasmid DNA. Biochimica et Biophysica Acta. 1991;**1088**:131-134. DOI: 10.1016/0167-4781(91)90162-f

[45] Aihara H, Miyazaki J. Gene transfer into muscle by electroporation *in vivo*. Nature Biotechnology. 1998;**16**:867-870. DOI: 10.1038/nbt0998-867

[46] Muramatsu T, Shibata O, Ryoki S, Ohmori Y, Okumura J. Foreign gene expression in the mouse testis by localized in vivo genetransfer. Biochemical and Biophysical Research Communications. 1997;**233**:45-49. DOI: 10.1006/bbrc.1997.6361

[47] Yamazaki Y, Fujimoto H, Ando H, Ohyama T, Hirota Y, et al. *In vivo* gene transfer to mouse spermatogenic cells by deoxyribonucleic acid injection into seminiferous tubules and subsequent electroporation. Biology of Reproduction. 1998;**59**:1439-1444. DOI: 10.1095/biolreprod59.6.1439

[48] Sato M, Ishikawa A, Kimura M. Direct injection of foreign DNA into mouse testis as a possible *in vivo* gene transfer system *via* epididymal spermatozoa. Molecular Reproduction and Development. 2002;**61**:49-56. DOI: 10.1002/mrd.1130

[49] Sato M, Tanigawa M, Kikuchi N, Nakamura S, Kimura M. Efficient gene delivery into murine ovarian cells by intraovarianinjection of plasmid DNA and subsequent *in vivo* electroporation. Genesis. 2003;**35**:169-174. DOI: 10.1002/gene.10182

[50] Saito T, Nakatsuji N. Efficient gene transfer into the embryonicmouse brain using *in vivo* electroporation. Developmental Biology. 2001;**240**:237-246. DOI: 10.1006/dbio.2001.0439

[51] Tabata H, Nakajima K. Efficient *in utero* gene transfer system to the developing mouse brain using electroporation: Visualization of neuronal migration in the developing cortex. Neuroscience. 2001;**103**:865-872. DOI: 10.1016/s0306-4522(01)00016-1

[52] Fukuchi-Shimogori T, Grove EA. Neocortex patterning by the secreted signaling molecule FGF8. Science. 2001;**294**:1071-1074. DOI: 10.1126/science.1064252

[53] Sato M, Ohtsuka M, Nakamura S. Intraoviductal instillation of a solution as an effective route for manipulating preimplantation mammalian embryos *in vivo*. In: Payan-Carreira R, editor. New Insights into Theriogenology. Rijeka (Croatia): InTechOpen; 2018. pp. 135-150

[54] Hakim BA, Tyagi V, Agnihotri SK, Nath A, Agrawal AK, et al. Electroporation of mouse follicles, oocytes and embryos without manipulating zona pellucida. Journal of Developmental Biology. 2021;**9**:13. DOI: 10.3390/jdb9020013

[55] Harrison MM, Jenkins BV, O'Connor-Giles KM, Wildonger JA. CRISPR view of development. Genes and Development. 2014;**28**:1859-1872. DOI: 10.1101/gad.248252.114

[56] Hsu PD, Lander ES, Zhang F. Development and applications of CRISPR-Cas9 for genome engineering. Cell. 2014;**157**:1262-1278. DOI: 10.1016/j.cell.2014.05.010

[57] Yoshimi K, Kunihiro Y, Kaneko T, Nagahora H, Voigt B, et al. ssODN-mediated knock-in with CRISPR-Cas for large genomic regions in zygotes. Nature Communications. 2016;**7**:10431. DOI: 10.1038/ncomms10431

[58] Chapman JR, Taylor MRG, Boulton SJ. Playing the end game: DNA double-strand break repair pathway choice. Molecular Cell. 2012;**47**:497-510. DOI: 10.1016/j.molcel.2012.07.029

[59] Orthwein A, Noordermeer SM, Wilson MD, Landry S, Enchev RI, et al. A mechanism for the suppression of homologous recombination in G1 cells. Nature. 2015;**528**:422-426. DOI: 10.1038/nature16142

[60] Wang H, Yang H, Shivalila CS, Dawlaty MM, Cheng AW, et al. One-step generation of mice carrying mutations in multiple genes by CRISPR/Cas-mediated genome engineering. Cell. 2013;**153**:910-918. DOI: 10.1016/j.cell.2013.04.025

[61] Yang H, Wang H, Shivalila CS, Cheng AW, Shi L, et al. One-step generation of mice carrying reporter and conditional alleles by CRISPR/Cas-mediated genome engineering. Cell. 2013;**154**:1370-1379. DOI: 10.1016/j.cell.2013.08.022

[62] Mashiko D, Fujihara Y, Satouh Y, Miyata H, Isotani A, et al. Generation of mutant mice by pronuclear injection of circular plasmid expressing Cas9 and single guided RNA. Scientific Reports. 2013;**3**:3355. DOI: 10.1038/srep03355

[63] Fujii W, Kawasaki K, Sugiura K, Naito K. Efficient generation of large-scale genome-modified mice using gRNA and CAS9 endonuclease. Nucleic Acids Research. 2013;**41**:e187. DOI: 10.1093/nar/gkt772

[64] Shen B, Zhang J, Wu H, Wang J, Ma K, et al. Generation of gene-modified mice via Cas9/RNA-mediated gene targeting. Cell Research. 2013;**23**:720-723. DOI: 10.1038/cr.2013.46

[65] Horii T, Arai Y, Yamazaki M, Morita S, Kimura M, et al. Validation of microinjection methods for generating knockout mice by CRISPR/Cas-mediated genome engineering. Scientific Reports. 2014;**4**:4513. DOI: 10.1038/srep04513

[66] Yen ST, Zhang M, Deng JM, Usman SJ, Smith CN, et al. Somatic mosaicism and allele complexity induced by CRISPR/Cas9 RNA injections in mouse zygotes. Developmental Biology. 2014;**393**:3-9. DOI: 10.1016/j.ydbio.2014.06.017

[67] Kaneko T, Sakuma T, Yamamoto T, Mashimo T. Simple knockout by electroporation of engineered endonucleases into intact rat embryos. Scientific Reports. 2014;**4**:6382. DOI: 10.1038/srep06382

[68] Sato M, Nakamura S, Inada E, Takabayashi S. Recent advances in the production of genome-edited rats. International Journal of Molecular Sciences. 2022;**23**:2548. DOI: 10.3390/ijms23052548

[69] Kaneko T, Mashimo T. Simple genome editing of rodent intact embryos by electroporation. PLoS One. 2015;**10**:e0142755. DOI: 10.1371/journal.pone.0142755

[70] Hashimoto M, Takemoto T. Electroporation enables the efficient mRNA delivery into the mouse zygotes and facilitates CRISPR/Cas9-based

genome editing. Scientific Reports. 2015;**5**:11315. DOI: 10.1038/srep11315

[71] Qin W, Dion SL, Kutny PM, Zhang Y, Cheng AW, et al. Efficient CRISPR/Cas9-mediated genome editing in mice by zygote electroporation of nuclease. Genetics. 2015;**200**:423-430. DOI: 10.1534/genetics.115.176594

[72] Chen S, Lee B, Lee AY, Modzelewski AJ, He L. Highly efficient mouse genome editing by CRISPR ribonucleoprotein electroporation of zygotes. The Journal of Biological Chemistry. 2016;**291**:14457-14467. DOI: 10.1074/jbc.M116.733154

[73] Tröder SE, Ebert LK, Butt L, Assenmacher S, Schermer B, et al. An optimized electroporation approach for efficient CRISPR/Cas9 genome editing in murine zygotes. PLoS One. 2018;**13**:e0196891. DOI: 10.1371/journal.pone.0196891

[74] Teixeira M, Py BF, Bosc C, Laubreton D, Moutin MJ, et al. Electroporation of mice zygotes with dual guide RNA/Cas9 complexes for simple and efficient cloning-free genome editing. Scientific Reports. 2018;**8**:474. DOI: 10.1038/s41598-017-18826-5

[75] Chen S, Sun S, Moonen D, Lee C, Lee AY, et al. CRISPR-READI: Efficient generation of knockin mice by CRISPR RNP electroporation and AAV donor infection. Cell Reports. 2019;**27**:3780-3789.e4. DOI: 10.1016/j.celrep.2019.05.103

[76] Remy S, Chenouard V, Tesson L, Usal C, Ménoret S, et al. Generation of gene-edited rats by delivery of CRISPR/Cas9 protein and donor DNA into intact zygotes using electroporation. Scientific Reports. 2017;**7**:16554. DOI: 10.1038/s41598-017-16328-y

[77] Miyasaka Y, Uno Y, Yoshimi K, Kunihiro Y, Yoshimura T, et al. CLICK: One-step generation of conditional knockout mice. BMC Genomics. 2018;**19**:318. DOI: 10.1186/s12864-018-4713-y

[78] Honda A, Tachibana R, Hamada K, Morita K, Mizuno N, et al. Efficient derivation of knock-out and knock-in rats using embryos obtained by *in vitro* fertilization. Scientific Reports. 2019;**9**:11571. DOI: 10.1038/s41598-019-47964-1

[79] Tanihara F, Takemoto T, Kitagawa E, Rao S, Do LT, et al. Somatic cell reprogramming-free generation of genetically modified pigs. Science Advances. 2016;**2**:e1600803. DOI: 10.1126/sciadv.1600803

[80] Tanihara F, Hirata M, Nguyen NT, Le QA, Wittayarat M, et al. Generation of CD163-edited pig via electroporation of the CRISPR/Cas9 system into porcine *in vitro*-fertilized zygotes. Animal Biotechnology. 2021;**32**:147-154. DOI: 10.1080/10495398.2019.1668801

[81] Sato M, Jin H, Akasaka E, Miyoshi M. *In vitro* electroporation in the presence of CRISPR/Cas9 reagents as a safe and useful method for producing biallelic knock out porcine embryos. OBM Genetics. 2021;**5**:15. DOI: 10.21926/obm.genet.2101123

[82] Kaneko T, Nakagawa Y. Genome editing of rodents by electroporation of CRISPR/Cas9 into frozen-warmed pronuclear-stage embryos. Cryobiology. 2020;**92**:231-234. DOI: 10.1016/j.cryobiol.2020.01.016

[83] Nakagawa Y, Kaneko T. Rapid and efficient production of genome-edited animals by electroporation into oocytes injected with frozen or freeze-dried sperm. Cryobiology. 2019;**90**:71-74. DOI: 10.1016/ j.cryobiol.2019.08.004

[84] Takahashi G, Gurumurthy CB, Wada K, Miura H, Sato M, Ohtsuka M. GONAD: *G*enome-editing via *O*viductal *N*ucleic *A*cids *D*elivery system: A novel microinjection independent genome engineering method in mice. Scientific Reports. 2015;**5**:11406. DOI: 10.1038/srep11406

[85] Ohtsuka M, Ogiwara S, Miura H, Mizutani A, Warita T, et al. Pronuclear injection-based mouse targeted transgenesis for reproducible and highly efficient transgene expression. Nucleic Acids Research. 2010;**38**:e198. DOI: 10.1093/nar/gkq860

[86] Ohtsuka M, Sato M, Miura H, Takabayashi S, Matsuyama M, et al. *i*-GONAD: A robust method for in situ germline genome engineering using CRISPR nucleases. Genome Biology. 2018;**19**:25. DOI: 10.1186/s13059-018-1400-x

[87] Kim S, Kim D, Cho SW, Kim J, Kim JS. Highly efficient RNA-guided genome editing in human cells via delivery of purified Cas9 ribonucleoproteins. Genome Research. 2014;**24**:1012-1019. DOI: 10.1101/gr.171322.113

[88] Quadros RM, Miura H, Harms DW, Akatsuka H, Sato T, et al. Easi-CRISPR: A robust method for one-step generation of mice carrying conditional and insertion alleles using long ssDNA donors and CRISPR ribonucleoproteins. Genome Biology. 2017;**18**:92. DOI: https://genomebiology.biomedcentral.com/articles/10.1186/s13059-017-1220-4

[89] Miura H, Quadros RM, Gurumurthy CB, Ohtsuka M. Easi-CRISPR for creating knock-in and conditional knockout mouse models using long ssDNA donors. Nature Protocols. 2018;**13**:195-215. Available from: https://www.nature.com/articles/nprot.2017.153

[90] Kobayashi T, Namba M, Koyano T, Fukushima M, Sato M, et al. Successful production of genome-edited rats by the rGONAD method. BMC Biotechnology. 2018;**18**:19. DOI: 10.1186/s12896-018-0430-5

[91] Takabayashi S, Aoshima T, Kabashima K, Aoto K, Ohtsuka M, et al. *i*-GONAD (improved genome-editing via oviductal nucleic acids delivery), a convenient *in vivo* tool to produce genome-edited rats. Scientific Reports. 2018;**8**:12059. DOI: 10.1038/s41598-018-30137-x

[92] Yoon Y, Wang D, Phillip WL, Tai PWL, Riley J, et al. Streamlined *ex vivo* and *in vivo* genome editing in mouse embryos using recombinant adeno-associated viruses. Nature Communications. 2018;**9**:412. DOI: 10.1038/s41467-017-02706-7

[93] Iwata S, Nakadai H, Fukushi D, Jose M, Nagahara M, et al. Simple and large-scale chromosomal engineering of mouse zygotes via *in vitro* and *in vivo* electroporation. Scientific Reports. 2019;**9**:14713. DOI: 10.1038/s41598-019-50900-y

[94] Hirose M, Ogura A. The golden (Syrian) hamster as a model for the study of reproductive biology: Past, present, and future. Reproductive Medicine and Biology. 2018;**18**:34-39. DOI: 10.1002/rmb2.12241

[95] Koyano T, Namba M, Kobayashi T, Nakakuni K, Nakano D, et al. The p21 dependent G2 arrest of the cell cycle in epithelial tubular cells links to the early stage of renal fibrosis. Scientific Reports. 2019;**9**:12059. DOI: 10.1038/s41598-019-48557-8

[96] Hirose M, Honda A, Fulka H, Tamura-Nakano M, Matoba S, et al. Acrosin is essential for sperm penetration through the zona pellucida in hamsters. Proceedings of the National Academy of Sciences of the United States of America. 2020;**117**:2513-2518. DOI: 10.1073/pnas.1917595117

[97] Sato M, Miyagasako R, Takabayashi S, Ohtsuka M, Hatada I, et al. Sequential *i*-GONAD: An improved *in vivo* technique for CRISPR/Cas9-based genetic manipulations in mice. Cell. 2020;**9**:546. DOI: 10.3390/cells9030546

[98] Kobayashi Y, Aoshima T, Ito R, Shinmura R, Ohtsuka M, et al. Modification of *i*-GONAD suitable for production of genome-edited C57BL/6 inbred mouse strain. Cell. 2020;**9**:957. DOI: 10.3390/cells9040957

[99] Takabayashi S, Aoshima T, Kobayashi Y, Takagi H, Akasaka E, et al. Successful *i*-GONAD in Brown Norway rats by modification of *in vivo* electroporation conditions. OBM Genetics. 2020;**4**(4):1-9. DOI: 10.21926/obm.genet.2004121

[100] Kaneko T, Tanaka S. Improvement of genome editing by electroporation using embryos artificially removed cumulus cells in the oviducts. Biochemical and Biophysical Research Communications. 2020;**527**:1039-1042. DOI: 10.1016/j.bbrc.2020.05.034

[101] Shang R, Zhang H, Bi P. Generation of mouse conditional knockout alleles in one step using the *i*-GONAD method. Genome Research. 2021;**31**:121-130. DOI: 10.1101/gr.265439.120

[102] Ferez M, Knudson CJ, Lev A, Wong EB, Alves-Peixoto P, et al. Viral infection modulates Qa-1b in infected and bystander cells to properly direct NK cell killing. The Journal of Experimental Medicine. 2021;**218**:e20201782. DOI: 10.1084/jem.20201782

[103] Nakano H, Kawai S, Ooki Y, Chiba T, Ishii C, et al. Functional validation of epitope-tagged ATF5 knock-in mice generated by improved genome editing of oviductal nucleic acid delivery (*i*-GONAD). Cell and Tissue Research. 2021;**385**:239-249. DOI: 10.1007/s00441-021-03450-7

[104] Yoshinaga S, Shin M, Kitazawa A, Ishii K, Tanuma M, et al. Comprehensive characterization of migration profiles of murine cerebral cortical neurons during development using FlashTag labeling. iScience. 2021;**24**:102277. DOI: 10.1016/j.isci.2021.102277

[105] Ho YT, Shimbo T, Wijaya E, Kitayama T, Takaki S, et al. Longitudinal single-cell transcriptomics reveals a role for Serpina3n-mediated resolution of inflammation in a mouse colitis model. Cellular and Molecular Gastroenterology and Hepatology. 2021;**12**:547-566. DOI: 10.1016/j.jcmgh.2021.04.004

[106] Umschweif G, Medrihan L, Guillén-Samander A, Wang W, Sagi Y, et al. Identification of Neurensin-2 as a novel modulator of emotional behavior. Molecular Psychiatry. 2021;**26**:2872-2885. DOI: 10.1038/s41380-021-01058-5

[107] Iwata S, Sasaki T, Nagahara M, Iwamoto T. An efficient *i*-GONAD method for creating and maintaining lethal mutant mice using an inversion balancer identified from the C3H/HeJJcl strain. G3 (Bethesda, Md.). 2021;**11**:jkab194. DOI: 10.1093/g3journal/jkab194

[108] Loubalova Z, Fulka H, Horvat F, Pasulka J, Malik R, et al. Formation of spermatogonia and fertile oocytes in golden hamsters requires piRNAs. Nature Cell Biology. 2021;**23**:992-1001. DOI: 10.1038/s41556-021-00746-2

[109] Miura H, Imafuku J, Kurosaki A, Sato M, Ma Y, et al. Novel reporter mouse models useful for evaluating *in vivo* gene editing and for optimization of methods of delivering genome editing tools. Molecular Therapy. Nucleic Acids. 2021;**24**:325-336. DOI: 10.1016/j.omtn.2021.03.003

[110] Zhang H, Shang R, Bi P. Feedback regulation of notch signaling and myogenesis connected by MyoD-Dll1 axis. PLoS Genetics. 2021;**17**:e1009729. DOI: 10.1371/journal.pgen.1009729

[111] Namba M, Kobayashi T, Kohno M, Koyano T, Hirose T, et al. Creation of X-linked Alport syndrome rat model with Col4a5 deficiency. Scientific Reports. 2021;**11**:20836. DOI: 10.1038/s41598-021-00354-y

[112] Aoshima T, Kobayashi Y, Takagi H, Iijima K, Sato M, et al. Modification of improved-genome editing via oviductal nucleic acids delivery (*i*-GONAD)-mediated knock-in in rats. BMC Biotechnology. 2021;**21**:63. DOI: 10.1186/s12896-021-00723-5

[113] Takaki S, Shimbo T, Ikegami K, Kitayama T, Yamamoto Y, et al. Generation of a recessive dystrophic epidermolysis bullosa mouse model with patient-derived compound heterozygous mutations. Laboratory Investigation. 2022;**102**:574-580. DOI: 10.1038/s41374-022-00735-5

[114] Hiradate Y, Harima R, Yanai R, Hara K, Nagasawa K, et al. Loss of Axdnd1 causes sterility due to impaired spermatid differentiation in mice. Reproductive Medicine and Biology. 2022;**21**:e12452. DOI: 10.1002/rmb2.12452

[115] Hasegawa A, Mochida K, Nakamura A, Miyagasako R, Ohtsuka M, et al. Use of anti-inhibin monoclonal antibody for increasing the litter size of mouse strains and its application to *i*-GONAD. Biology of Reproduction. 2022;**107**:605-618. DOI: 10.1093/biolre/ioac068

[116] Gurumurthy CB, Takahashi G, Wada K, Miura H, Sato M, et al. GONAD: A novel CRISPR/Cas9 genome editing method that does not require *ex vivo* handling of embryos. Current Protocols in Human Genetics. 2016;**88**:15.8.1-15.8.12. DOI: 10.1002/0471142905.hg1508s88

[117] Gurumurthy CB, Sato M, Nakamura A, Inui M, Kawano N, et al. Creation of CRISPR-based germline-genome-engineered mice without *ex vivo* handling of zygotes by *i*-GONAD. Nature Protocols. 2019;**14**:2452-2482. DOI: 10.1038/s41596-019-0187-x

[118] Ohtsuka M, Sato M. *i*-GONAD: A method for generating genome-edited animals without *ex vivo* handling of embryos. Development Growth and Differentiation. 2019;**61**:306-315. DOI: 10.1111/dgd.12620

[119] Namba M, Kobayashi T, Koyano T, Kohno M, Ohtsuka M, et al. GONAD: A new method for germline genome-editing in mice and rats. Development Growth and Differentiation. 2021;**63**:439-447. DOI: 10.1111/dgd.12746

[120] Kjell J, Olson L. Rat models of spinal cord injury: From pathology to potential therapies. Disease Models and Mechanisms. 2016;**9**:1125-1137. DOI: 10.1242/dmm.025833

[121] Jacob HJ. The rat: A model used in biomedical research. In: Anegon I, editor. Rat Genomics: Methods and Protocols Methods in Molecular Biology. Totowa, NJ (USA): Humana Press; 2010. pp. 1-11. DOI: 10.1007/978-1-60327-389-3_1

[122] Geurts AM, Cost GJ, Freyvert Y, Zeitler B, Miller JC, et al. Knockout rats

via embryo microinjection of zinc-finger nucleases. Science. 2009;**325**:433. DOI: 10.1126/science.1172447

[123] Zhang L, Shao Y, Li L, Tian F, Cen J, et al. Efficient liver repopulation of transplanted hepatocyte prevents cirrhosis in a rat model of hereditary tyrosinemia type I. Scientific Reports. 2016;**6**:31460. DOI: 10.1038/srep31460

[124] Nakamura K, Fujii W, Tsuboi M, Tanihata J, Teramoto N, et al. Generation of muscular dystrophy model rats with a CRISPR/Cas system. Scientific Reports. 2014;**4**:5635. DOI: 10.1038/srep05635

[125] Larcher T, Lafoux A, Tesson L, Remy S, Thepenier V, et al. Characterization of dystrophin deficient rats: A new model for Duchenne muscular dystrophy. PLoS One. 2014;**9**:e110371. DOI: 10.1371/journal.pone.0110371

[126] Whittaker D. "Hamster" in the UFAW handbook on the care and Management of Laboratory Animals. In: Trevor P, editor. Oxford. London (United Kingdom): Blackwell Science Ltd.; 1999. pp. 356-366

[127] Yanagimachi R, Yanagimachi H, Rogers BJ. The use of zona-free animal ova as a test-system for the assessment of the fertilizing capacity of human spermatozoa. Biology of Reproduction. 1976;**15**:471-476. DOI: 10.1095/biolreprod15.4.471

[128] Schini SA, Bavister BD. Two-cell block to development of cultured hamster embryos is caused by phosphate and glucose. Biology of Reproduction. 1988;**39**:1183-1192. DOI: 10.1095/biolreprod39.5.1183

[129] Galili U. The α-gal epitope (gal α1-3Gal β1-4GlcNAc-R) in xenotransplantation. Biochimie. 2001;**83**:557-563. DOI: 10.1016/S0300-9084(01)01294-9

[130] Horii T, Morita S, Kimura M, Terawaki N, Shibutani M, et al. Efficient generation of conditional knockout mice via sequential introduction of *lox* sites. Scientific Reports. 2017;**7**:7891. DOI: 10.1038/s41598-017-08496-8

[131] Lee AY, Lloyd KCK. Conditional targeting of Ispd using paired Cas9 nickase and a single DNA template in mice. FEBS Open Bio. 2014;**4**:637-642. DOI: 10.1016/j.fob.2014.06.007

[132] Bishop KA, Harrington A, Kouranova E, Weinstein EJ, Rosen CJ, et al. CRISPR/Cas9-mediated insertion of *loxP* sites in the mouse Dock7 gene provides an effective alternative to use of targeted embryonic stem cells. G3 (Bethesda). 2016;**6**:2051-2061. DOI: 10.1534/g3.116.030601

[133] Nakagawa Y, Oikawa F, Mizuno S, Ohno H, Yagishita Y, et al. Hyperlipidemia and hepatitis in liver-specific CREB3L3 knockout mice generated using a one-step CRISPR/Cas9 system. Scientific Reports. 2016;**6**:27857. DOI: 10.1038/srep27857

[134] Ma X, Chen C, Veevers J, Zhou XM, Ross RS, et al. CRISPR/Cas9-mediated gene manipulation to create single-amino-acid-substituted and foxed mice with a cloning-free method. Scientific Reports. 2017;7:42244. DOI: 10.1038/srep42244

[135] Sung YK, Kim SW. Recent advances in the development of gene delivery systems. Biomaterials Research. 2019;**23**:8. DOI: 10.1186/s40824-019-0156-z

[136] Ellis BL, Hirsch ML, Barker JC, Connelly JP, Steininger RJ III, et al. A survey of *ex vivo/in vitro* transduction efficiency of mammalian primary cells and cell lines with nine natural adeno-associated virus (AAV1-9) and

one engineered adeno-associated virus serotype. Virology Journal. 2013;**10**:74. DOI: 10.1186/1743-422X-10-74

[137] Sonntag F, Schmidt K, Kleinschmidt JA. A viral assembly factor promotes AAV2 capsid formation in the nucleolus. Proceedings of the National Academy of Sciences of the United States of America. 2010;**107**:10220-10225. DOI: 10.1073/pnas.1001673107

[138] Sato M, Takabayashi S, Akasaka E, Nakamura S. Recent advances and future perspectives of *in vivo* targeted delivery of genome-editing reagents to germ cells, embryos, and fetuses in mice. Cell. 2020;**9**:799. DOI: 10.3390/cells9040799

[139] Sato M, Sato-Yamamoto N, Wakita A, Haraguchi M, Shimonishi M, et al. Direct injection of recombinant AAV-containing solution into the oviductal lumen of pregnant mice caused in situ infection of both preimplantation embryos and oviductal epithelium. International Journal of Molecular Sciences. 2022;**23**:4897. DOI: 10.3390/ijms23094897

[140] Maruyama T, Dougan SK, Truttmann MC, Bilate AM, Ingram JR, et al. Increasing the efficiency of precise genome editing with CRISPR-Cas9 by inhibition of nonhomologous end joining. Nature Biotechnology. 2015;**33**:538-542. DOI: 10.1038/nbt.3190

[141] Ma Y, Chen W, Zhang X, Yu L, Dong W, et al. Increasing the efficiency of CRISPR/Cas9-mediated precise genome editing in rats by inhibiting NHEJ and using Cas9 protein. RNA Biology. 2016;**13**:605-612. DOI: 10.1080/15476286.2016.1185591

[142] Song J, Yang D, Xu J, Zhu T, Chen YE, et al. RS-1 enhances CRISPR/Cas9- and TALEN-mediated knock-in efficiency. Nature Communications. 2016;**7**:10548. DOI: 10.1038/ncomms10548

[143] Li G, Zhang X, Zhong C, Mo J, Quan R, et al. Small molecules enhance CRISPR/Cas9-mediated homology-directed genome editing in primary cells. Scientific Reports. 2017;**7**:8943. DOI: 10.1038/s41598-017-09306-x

[144] Walton EA, Evans G, Armstrong DT. Ovulation response and fertilization failure in immature rats induced to superovulate. Journal of Reproduction and Fertility. 1983;**67**:91-96. DOI: 10.1530/jrf.0.0670091

[145] Van der Auwera I, Pijnenborg R, Koninckx PR. The influence of *in-vitro* culture versus stimulated and untreated oviductal environment on mouse embryo development and implantation. Human Reproduction. 1999;**14**:2570-2574. DOI: 10.1093/humrep/14.10.2570

[146] Park SJ, Kim TS, Kim JM, Chang KT, Lee HS, et al. Repeated superovulation via PMSG/ hCG administration induces 2-Cys peroxiredoxins expression and overoxidation in the reproductive tracts of female mice. Molecules and Cells. 2015;**38**:1071-1078. DOI: 10.14348/molcells.2015.0173

[147] Christenson CM, Eleftheriou BE. Dose-dependence of superovulation response in mice to two injections of PMSG. Journal of Reproduction and Fertility. 1972;**29**:287-289. DOI: 10.1530/jrf.0.0290287

[148] Fox JG, Barthold S, Davisson M, Newcomer CE, Quimby FW, et al. The mouse in biomedical research: Normative biology, husbandry, and models. In: Fox J, editor. Kindle. 2nd ed. Vol. 3. London (United Kingdom): Elsevier; 2006

[149] Teng K, Ford MJ, Harwalkar K, Li Y, Pacis AS, et al. Modeling high-grade serous ovarian carcinoma using a combination of *in vivo* fallopian tube electroporation and CRISPR-Cas9–mediated genome editing. Cancer Research. 2021;**81**:5147-5160. DOI: 10.1158/0008-5472.CAN-20-1518

[150] Hryhorowicz M, Lipiński D, Hryhorowicz S, Nowak-Terpiłowska A, Ryczek N, et al. Application of genetically engineered pigs in biomedical research. Genes (Basel). 2020;**11**:670. DOI: 10.3390/genes11060670

[151] Whyte JJ, Prather RS. Genetic modifications of pigs for medicine and agriculture. Molecular Reproduction and Development. 2011;**78**:879-891. DOI: 10.1002/mrd.21333

[152] Wu Y, Zhang J, Peng B, Tian D, Zhang D, et al. Generating viable mice with heritable embryonically lethal mutations using the CRISPR-Cas9 system in two-cell embryos. Nature Communications. 2019;**10**:1-13. DOI: 10.1038/s41467-019-10748-2

[153] Anzalone AV, Randolph PB, Davis JR, Sousa AA, Koblan LW, et al. Search-and-replace genome editing without double-strand breaks or donor DNA. Nature. 2019;**576**:149-157. DOI: 10.1038/s41586-019-1711-4

[154] Chen PJ, Hussmann JA, Yan J, Knipping F, Ravisankar P, et al. Enhanced prime editing systems by manipulating cellular determinants of editing outcomes. Cell. 2021;**184**:5635-5652.e29. DOI: 10.1016/j.cell.2021.09.018

[155] Anzalone AV, Gao XD, Podracky CJ, Nelson AT, Koblan LW, et al. Programmable deletion, replacement, integration and inversion of large DNA sequences with twin prime editing. Nature Biotechnology. 2022;**40**:731-740. DOI: 10.1038/s41587-021-01133-w

[156] Li X, Zhou L, Gao B-Q, Li G, Wang X, et al. Highly efficient prime editing by introducing same-sense mutations in pegRNA or stabilizing its structure. Nature Communications. 2022;**13**:1669. DOI: 10.1038/s41467-022-29339-9

[157] Komor AC, Kim YB, Packer MS, Zuris JA, Liu DR. Programmable editing of a target base in genomic DNA without double-stranded DNA cleavage. Nature. 2016;**533**:420-424. DOI: 10.1038/nature17946

[158] Nishida K, Arazoe T, Yachie N, Banno S, Kakimoto M, et al. Targeted nucleotide editing using hybrid prokaryotic and vertebrate adaptive immune systems. Science. 2016;**353**:aaf8729. DOI: 10.1126/science.aaf8729

[159] Ma Y, Zhang J, Yin W, Zhang Z, Song Y, et al. Targeted AID-mediated mutagenesis (TAM) enables efficient genomic diversification in mammalian cells. Nature Methods. 2016;**13**:1029-1035. DOI: 10.1038/nmeth.4027

Section 3

Application

Chapter 6

Applications of CRISPR/Cas9 for Selective Sequencing and Clinical Diagnostics

Maximilian Evers, Björn Brändl, Franz-Josef Müller, Sönke Friedrichsen and Stephan Kolkenbrock

Abstract

In this chapter, we will discuss the applications of CRISPR/Cas9 in the context of clinical diagnostics. We will provide an overview of existing methods and their use cases in the diagnostic field. Special attention will be given to selective sequencing approaches using third-generation sequencing and PAM-site requirements. As target sequences in an AT-rich environment cannot easily be accessed by the commercially available SpCas9 due to rarity of NGG PAM-sites, new enzymes such as ScCas9 with PAM-site requirements of NNG will be highlighted. Original research on CRISPR/Cas9 systems to determine molecular glioma markers by enriching regions of interest will be discussed in the context of potential future applications in clinical diagnostics.

Keywords: CRISPR/Cas9, clinical diagnostics, selective sequencing, PAM site, cancer

1. Introduction

1.1 Current diagnostic context

Emerging infectious diseases such as COVID-19, acquired or hereditary genetic defects causing cancer and other illnesses fuel the need for fast and cost-effective diagnostic tools. Gold standard for many types of disease detection is the real-time polymerase chain reaction (PCR) due to its robustness and sensitivity toward molecular biomarkers associated with diseases. Especially, quantitative PCR (qPCR) and reverse transcriptase qPCR have been staples of infectious disease diagnostics to determine the presence of pathogens and viral loads [1, 2], but have also proven their efficacy in tumor diagnostics due to high sensitivity and low input requirements [3, 4]. Many next-generation sequencing (NGS) approaches are also used as diagnostic tools. Whole exome and genome sequencing is used for the investigation of many molecular markers. The benefits of NGS include the ability to screen a large amount of possible target genes in tandem for comparatively low cost [5]. Furthermore, NGS can be used for unbiased detection and species level determination of pathogens in septic patients. This removes the need for time-consuming blood cultures [6]. Other well-established methods in diagnostics are based on antigen-antibody interaction, such as the Enzyme-Linked Immunosorbent Assay (ELISA) or paper-based lateral flow assays [7, 8].

A new addition to the diagnostic toolbox is clustered regularly interspaced short palindromic repeats (CRISPR)-based diagnostics. CRISPR-associated (Cas) proteins are RNA-guided endonucleases originally part of the adaptive immune system of prokaryotes to ward off invading nucleic acids. Several types of CRISPR/Cas systems have been discovered, and some have been used for diagnostic applications such as Cas12 and Cas13 for methods such as DNA endonuclease-targeted CRISPR *trans* reporter (DETECTR) and specific high-sensitivity enzymatic reporter unlocking (SHERLOCK) and SHERLOCKv2 [9], which were recently developed as potent tools for COVID-19 detection. This chapter will focus on the utility of CRISPR/Cas9 in clinical diagnostics.

1.2 CRISPR/Cas9

CRISPR RNAs (crRNA) provide the targeting mechanism for the Cas9 nuclease activity. crRNAs are hybridized with trans-activating crRNA (tracrRNA), providing a stem-loop structure that anchors the RNA-complex to the Cas9 protein. The crRNA can be engineered to target a wide array of sequences rendering CRISPR/Cas9 a powerful tool for targeted gene editing and recognition. Cas9 proteins are characterized by two nuclease domains forming the active center, HNH and RuvC [10]. HNH is a single nuclease domain responsible for cleaving the DNA strand complementary to the RNA guide. RuvC is split into three subdomains, with RuvC I at the N-terminus of the protein and RuvC II/III flanking the HNH domain near the center of the amino acid sequence [11]. The catalytic residues D10 (in RuvC I) and H840 (in HNH) can be substituted to either limit nuclease activity in case of a single-site inactivation to create a Cas9 nickase or to generate a catalytically inactive/dead Cas9 (dCas9) variant in case of a double-site inactivation [12]. In addition to the nuclease domains, Cas9 possesses a protospacer adjacent motif (PAM)-interacting (PI) domain. The PAM is a short nucleic acid sequence downstream of the crRNA conferred target sequence, required for nuclease activity and target sequence interrogation [13]. It is thought to have originated in prokaryotes so as not to target their own DNA and thus to prevent an autoimmune response [14]. While the sequence to be cut can be easily defined via crRNA, the obligatory requirement of a protospacer adjacent motif (PAM) sequence next to the target sequence [15, 16] limits the applications of Cas9 in clinical diagnostics. Due to this limitation, regions of interest without matching PAM-site cannot be cleaved and subsequently analyzed. Several variants of Cas9 enzymes have been generated to partially circumvent those limitations with a relaxation of the PAM-site requirement. The Cas9 from *Streptococcus pyogenes* (SpCas9) natively recognizes the PAM 5'-NGG-3' but was modified (termed xCas9) to accept a broad range of PAM sites, including 5'-NG-3', 5'-GAA-3', and 5'-GAT-3' [17]. Additionally, Cas9 enzymes from different hosts such as the Cas9 from *Streptococcus canis* (ScCas9) were modified to be more promiscuous regarding PAM site recognition (termed Cas9-SC++), now accepting 5'-NNG-3' as a PAM site [18]. A Cas9 homolog discovered in *Francisella novicida* (FnCas9) also recognizes the 5'-NGG-3' PAM but was successfully engineered to accept a 5'-YG-3' PAM [19].

1.3 Cas9 in diagnostic methods

In the area of molecular diagnostics, CRISPR/Cas9 systems have proven to be effective tools in distinguishing between different Zika virus strains. Pardee et al. (2016) [20] used nucleic acid sequence-based amplification (NASBA) in combination with Zika strain-specific sgRNA/Cas9 and toehold switches to create a colorimetric assay to detect and differentiate African and American Zika virus strains. A toehold switch is an RNA molecule combining a sensor and a reporter sequence. Without the presence of the trigger, an RNA molecule complementary to the sensor sequence, a hairpin structure is formed. It limits access to the ribosomal binding site and therefore inhibits translation of the reporter. Due to strand displacement upon hybridization with the trigger RNA, the hairpin structure is resolved, allowing the translation of the reporter [21]. The toehold switch was designed to regulate *lacZ* expression and was activated by the Zika virus RNA amplicons, which allowed for colorimetric *in vitro* detection of the target RNA. Due to sequence differences, PAM site locations vary between the strains, which was exploited for targeted truncation of RNA amplicons of only one strain in a method termed NASBA-CRISPR Cleavage (NASBACC). Truncated RNA amplicons could not activate the toehold switch, which allowed for discrimination between the strains [20].

CRISPR-Cas9 nickase (SpCas9H840A nickase) strand displacement amplification (CRISDA) is an ultrasensitive method to detect target DNA with single-nucleotide accuracy and attomolar sensitivity. A pair of SpCas9 nickase ribonucleoproteins (RNPs) introduce nicks in the flanking areas next to the region of interest. Initial primers anneal to the nicked strands from where strand displacement amplification begins. Biotin and Cy5-labeled peptide nucleic acid (PNAs) probes are introduced to the amplification mix to detect and quantify amplicons. The PNA binds to the amplicons and enables a pulldown using streptavidin-coated magnetic beads. Fluorescence measurement of the pulled-down DNA allows for quantification of the generated amplicons [22].

Another nucleic acid detection strategy is CRISPR/Cas9-triggered isothermal exponential amplification reaction (CAS-EXPAR). It is based on CRISPR/Cas9 cleavage and nicking endonuclease (NEase)-mediated nucleic acids amplification. Cas9 cleavage of the target produces a primer for the CAS-EXPAR reaction, wherein the target "X" hybridizes with a construct containing two sequences complementary to the target ("X'"), which are connected via a PAMmer. Upon extension of the double strand, Cas9 cleaves off the newly synthesized DNA, which in turn acts as a primer itself. This strategy was shown to have a detection limit of 0.82 amol and high specificity, discriminating single-base mismatches [23].

Lateral flow assays are state of the art in point-of-care diagnostics. CRISPR/Cas9-mediated lateral flow nucleic acid assay (CASLFA) combines the sensitivity of Cas9 endonuclease with the ease of use of lateral flow assays. CASLFA was developed for the identification of *Listeria monocytogenes*, different genetically modified organisms (GMOs), and the African swine fever virus (ASFV) [24].

FnCas9 editor-linked uniform detection assay (FELUDA) is a diagnostic tool combining preamplification of a target sequence using biotinylated primers with inactive FnCas9 to detect target sequences. The used tracrRNA is FAM-labeled and can be recognized via antibodies, and the capture of target sequences is paired with a lateral flow readout. The biotinylated amplicons bind to the test region via streptavidin interaction. If FnCas9 binds to the amplicon DNA, it will be retained in the test region, allowing for antibody-based detection in the form of a visible band. dFnCas9 was used for this assay, as it exhibits lower affinity toward sequences with single-nucleotide mismatches to the crRNA used than SpCas9 [25].

Finding Low Abundance Sequences by Hybridization (FLASH) is a method that combines Cas9 digestion, PCR, and Illumina sequencing to detect and identify antimicrobial resistance in microbial DNA samples. Isolated DNA is dephosphorylated before Cas9 digestion of target sequences. The double-strand breaks introduced by Cas9 remain phosphorylated and are subsequently dA-tailed. Adapters are ligated to the dA-tailed target sequences, which are then amplified via PCR. The resulting library can be sequenced via Illumina sequencing and achieve sub-attomolar sensitivity [26].

Next to infectious disease detection, another field of interest for targeted Cas9 diagnostics is cancer, one of the world's leading causes of premature death [27]. As cancerous unregulated cell growth can be caused by a combination of genetic defects, it is vital for prognosis and treatment to accurately diagnose its molecular cause. Cancer diagnostics currently is often based on histological analysis of tumor tissue. Because histology is predetermined by genetics, research efforts to quickly identify aberrant tumor marker genes on the molecular level are being pursued.

One application to potentially target this challenge is CRISPR-Chip. This method utilizes dCas9 immobilized on a graphene surface, acting as a conductor. The dCas9 is paired with a specifically designed sgRNA to recognize its target. Upon binding target DNA, the conductivity of the immobilized dCas9 changes, which can be measured via the graphene surface. This allows for detection limits of 1.7 fM gDNA. Though it was demonstrated with target sequences associated with Duchenne muscular dystrophy, it could be used for any sequences as long as a suitable PAM-site is flanking the region of interest [28].

Another route to follow in molecular tumor diagnosis is the sequencing of tumor marker genes. With the advent of second- and third-generation sequencing, the feasibility of sequencing approaches in standard diagnostics is increasing due to lower costs and shorter sequencing times. However, a combination with CRISPR/Cas technology allows for a specific sequencing of the regions of interest, boosting the output of relevant regions, and thus enabling a faster and very specific and sensitive sequencing approach.

1.4 Tumor biomarker selection for Cas9-targeted sequencing

To maximize utility of Cas9-targeted sequencing, biomarkers such as mutations or methylation patterns with defined locations are favorable. Because sequencing times are determined by target sequence length, biomarkers such as defined SNPs allow for higher throughput, as flanks of the targeted sequence can be chosen in close proximity to the region of interest. In our research we developed an amplification-independent workflow to assess the tumor marker status of six relevant genes/regions in brain tumors following the 2021 WHO Classification of Tumors of the Central Nervous System [29]. These genes/regions, their function, and glioma-relevant mutations are described in the following.

Isocitrate dehydrogenases 1 and 2 (IDH1, IDH2) are crucial enzymes that catalyze the oxidative decarboxylation of isocitrate to α-ketoglutarate during the Krebs cycle. Common mutations associated with glioma formation are related to codons 132 for IDH1 and 172 for IDH2 causing aberrant enzymatic activity and in turn the accumulation of 2-hydroxyglutarate, which inhibits many α-ketoglutarate dependent enzymes such as DNA-demethylases, leading to DNA hypermethylation [30].

Additionally, the promoter of telomerase reverse transcriptase (pTERT) represents a clinically relevant target due to its close association with oncogenesis and immortalization of cell lines [31]. The mutations C228T and C250T are commonly associated with aberrant expression patterns as these mutations create *de novo* binding sites for members of the E26 transformation-specific family of transcription factors [31].

H3F3A and *Hist1H3B* encode histone subunits H3.3 and H3.1, respectively. K27M variants are observed in different cancer types, such as Diffuse Intrinsic Pontine Glioma, and G34R/V substitution in H3.3 is also associated with young adult high-grade astrocytoma [32, 33].

BRAF encodes a member of the Raf kinase family, B-Raf, and is a growth signal transduction protein kinase that regulates pathways associated with cell division and differentiation. The amino acid substitution V600E of B-Raf increases its basal activity and stimulates cell division and differentiation pathways. This is associated with a variety of different cancer types [34].

As mutation detection via sequencing is of interest, it is crucial to be aware of the benefits and drawbacks of the used sequencing technologies for the development of a medical diagnostic application.

1.5 Current sequencing technologies

Currently, the most widely used next-generation sequencing technologies on the market are Illumina short-read sequencing, PacBio (also referred to as Single-Molecule Real-Time (SMRT) sequencing), 454 pyrosequencing, ion-torrent/proton sequencing, and nanopore sequencing by Oxford Nanopore Technologies (ONT) [35]. Illumina, 454, and ion-torrent sequencing technologies are referred to as second-generation sequencing technologies. They deliver short reads of about 50–1000 bp in length and their parallelization in sequencing reaction results in a high read throughput (0.7–15 M reads per run) and an amount of sequence information of about 0.5–8.5 Gb per run [36]. PacBio and nanopore sequencing are referred to as third-generation sequencing technologies. They usually deliver tens of kb per read up to several Mb, but far fewer reads in total. For nanopore sequencing, the amount of sequence information is strongly dependent on the flow cell. Depending on the nucleic acid library preparation and its quality, up to 2.8 Gb per run is theoretically achievable on a Flongle flow cell, 10–15 Gb on a MinION flow cell [37], and up to 153 Gb per run was reported by using the PromethION flow cell [38]. PacBio sequencing utilizes SMRT cells for sequencing, usually generating 55,000–365,000 reads per run with an average read length of 10–16 kb [39] and 15–96 Gb per run [40].

Nanopore and PacBio sequencing allow for real-time sequencing with parallel base calling of the steadily increasing raw sequencing information allowing direct usage of the results during the run. In addition, both techniques allow detection of epigenetic information of each nucleotide sequenced [41], which can be a piece of important additional information in clinical cancer diagnostics and treatment [42–44]. While methylation of a base directly impacts the raw signal of the nanopore sequencing and thus can be distinguished from an unmodified nucleotide, PacBio detects methylation by a change in DNA-polymerase kinetics during synthesis. Due to the "sequencing by synthesis" technology of second-generation sequencing techniques, they cannot detect epigenetic modification directly, but only via a pretreatment step such as bisulfite treatment, endonuclease digestion, or affinity enrichment [45].

In summary, the selection of the sequencing technology used for clinical diagnostics will be strongly dependent on requirements such as mode of analysis and time to

Sequencing technology	Real-time	Average sequence length	Typical number of reads	Amount sequence data per run	Methylation status detection
Illumina	no	2 × 150 bp (HiSeq 4000) [39]	5 billion (HiSeq 4000) [39]	1300–1500 Gb*	no
		2 × 300 bp (MiSeq) [39]	25 million (MiSeq) [39]	4.5–5.1 Gb*	
454 pyro-sequencing	no	400–1000 bp [46]	> 1 million [32]	35–450 Mb [46]	no
Ion-torrent	no	200–600 bp*	2–130 million*	0.3–50 Gb*	no
PacBio	yes	10–16 kb [39]	55,000–365,000 [25]	15 Gb (Sequel) [40]	yes
				96 Gb (Sequel II) [40]	
Oxford Nanopore	yes	10–30 kb (all flow cells) record 2.6 Mb [47]	200,000 (Flongle) [48]	Up to 2.8 Gb (Flongle)*	yes
			1,200,000 (MinION) [48]	10–15 Gb (MinION) [37]	
			Up to 250 million (PromethION)*	Up to 153 Gb (PromethION) [38]	

**Specifications obtained from the respective company´s website (as of 01.06.2022).*

Table 1.
Most widely used sequencing technologies and their characteristics.

results. The characteristics in this respect of each mentioned sequencing technology are summarized in **Table 1**.

Next to Illumina sequencing-based methods such as FLASH, third-generation sequencing can also be paired with Cas9 enrichment of target sequences. PacBio uses a generic SMRT sequencing library, which is digested by Cas9. The digested sequences are then ligated to a second set of adapters, which is used for magnetic-bead-based separation of the targeted sequences, allowing for target sequence enrichment [49].

For ONT's Cas9-targeted sequencing process (nCATS), DNA is dephosphorylated before Cas9 digestion. Like FLASH, the phosphorylated ends of the cleaved DNA are dA-tailed and ligated to sequencing adapters, allowing for selective sequencing [50]. A similar approach was pursued in the following experimental section to develop a CRISPR/Cas, third-generation sequencing assay for diagnosis of. Utilizing the promising properties of nanopore and Cas9-dependent target enrichment, we developed an amplification-independent workflow to assess glioma biomarkers.

2. Development of a CRISPR/Cas9-targeted sequencing approach

2.1 Material and methods

To test the feasibility of nanopore sequencing in brain tumor marker detection, we used pUC57 vectors containing 2 kb target sequence of a given tumor marker as either wild-type or containing a clinically relevant mutation. Cas9-RNP populations were prepared to cleave the DNA upstream and downstream of a given mutation site. The excised double-stranded DNA was used for sequencing library preparation using the SQK-CS9109 Cas9 sequencing Kit from ONT. Flongle flow cells (version R.9.4.1) were used for sequencing. Sequences were assessed using a minimap2 [51] alignment followed by custom SNP calling using python scripts. As tumor treatment is very time-sensitive [52], the possibility of intra-surgical diagnostics could alleviate an unmet clinical need. Therefore, we evaluated the results not only by accuracy but also regarding time to results. In addition to the complete sequencing data acquired a subset generated during the first 15 min of sequencing was also used for analysis.

2.2 Cas9-RNP preparation

Alt-R® S.p. Cas9 Nuclease V3 including tracrRNA and crRNAs were purchased from Integrated DNA Technologies (Coralville, IA, USA). The crRNAs were designed to target at least 200 bp upstream and downstream of each mutation site resulting in at least 1000 bp of excised dsDNA in total. crRNAs were designed for *IDH1, IDH2,* pTERT, *H3F3A, Hist1H3B,* and *BRAF.* All sequences of used crRNAs are given in **Table 2**. To anneal crRNA and tracrRNA, 8 μL Duplex Buffer (IDT), 1 μL tracrRNA (100 μM), and 1 μL crRNA Pool (100 μM, equimolar) were assembled in 0.2 mL thin-walled PCR tubes and incubated at 95 °C in a thermal cycler. The mix was allowed to cool to room temperature (RT) afterward. The annealed crRNA/tracrRNA (10 μM) was added to 79.2 μL nuclease-free water, 10 μL Reaction buffer (SQK-CS9109 Kit), and 0.8 μL Alt-R® S.p. Cas9 Nuclease V3 (62 μM) and mixed thoroughly by flicking. The RNPs were formed by incubation at RT for 30 min and stored at 4°C until needed. Two different RNP populations were prepared with different crRNAs. Population 1 included all crRNAs described in **Table 2**, whereas population 2 was prepared with only the two crRNAs targeting *IDH1*.

2.3 Cas9 digestion of pUC57 plasmids

The pUC57 vectors containing tumor marker sequences used as target DNA for Cas9 digestion were purchased from GenScript (Piscataway, NJ, USA). For DNA digestion and library preparation, the SQK-CS9109 Cas9 sequencing kit from ONT (Oxford, UK). DNA digestion was set up by adding template plasmids to Cas9 RNPs, reaction buffer, dATP, and Taq DNA-polymerase. One digestion was performed with

Target	crRNA Seq fw (5'-3')	crRNA Seq rv (5'-3')	Fragment size [bp]
IDH1	ATGTTTAATACAATCTTTGG	GCTTCCCATTACAAGAGGAG	1062
IDH2	AGTGCACACGATGTTTCTGC	TCGTCCTCACGACAACACTT	1601
pTERT	CTCCCTGACGCTATGGTTCC	GTCAAGGAGCCCAAGTCGCG	1443
H3F3A	AATTTGACTCGACCTTCCAG	TATTTGCGGAGGCTAAGTCT	1081
Hist1H3B	GCATTCCTAACTATCTTGAA	CATAGTCTAATGCTTTCCGG	1334
BRAF	GACCCTCTAAAACGGTGTGA	GCATGCATGTATAGGAGAGC	1584

Table 2.
List of crRNAs used for enzymatic excision of tumor marker DNA from pUC57 vectors containing target gene fragments.

	Digestion 1	Digestion 2
pUC57::IDH1_Wt (80 ng/μL)	—	2 μL
pUC57::IDH1_R132C (80 ng/μL)	1 μL	—
pUC57::IDH1_R132H (80 ng/μL)	—	2 μL
pUC57::IDH2_Wt (80 ng/μL)	1 μL	—
pUC57::pTERT_Wt (80 ng/μL)	1 μL	—
pUC57::H3F3A_Wt (80 ng/μL)	1 μL	—
pUC57::Hist1H3B_Wt (80 ng/μL)	1 μL	—
pUC57::BRAF_Wt (80 ng/μL)	1 μL	—
ddH2O	21 μL	23 μL
Cas9 RNPs (Population 1) (10 μM)	10 μL	—
Cas9 RNPs (Population 2) (10 μM)	—	10 μL
Reaction Buffer	3 μL	3 μL
dATP	1 μL	1 μL
Taq Polymerase	1 μL	1 μL

Table 3.
Setup for digestion of different pUC57 plasmid mixtures using different Cas9-RNP populations.

a mixture of six plasmids (80 ng each), each containing a different marker, in order to test multiplexing. Another digestion was set up with a mixture of two plasmids (160 ng each) containing different mutations of the same marker, *IDH1*, in order to test variant calling capabilities. The different reaction mixtures were prepared as shown in **Table 3**.

The reaction mixtures were incubated for 30 min at 37°C for Cas9 cleavage. Subsequently it was incubated at 72°C for 5 min for Taq Polymerase facilitated dA-tailing of cleaved fragments.

2.4 Library preparation

For sequencing adapter ligation to the dA-tailed fragments, the digested and dA-tailed DNA was added to a mixture of 20 μL Ligation Buffer, 3 μL nuclease-free water, 10 μL T4 Ligase, and 5 μL Adapter Mix. The ligation components were mixed by flicking, spun down, and incubated at RT for 10 min. DNA was purified using AMPure XP beads (Beckman Coulter, Brea, CA, USA). In total, 48 μL of magnetic beads was added to each sample and mixed by inversion. The samples were incubated for 10 min at RT without agitation. Afterward, samples were spun down, and beads were separated magnetically. The supernatant was discarded, and the pellet washed twice with 250 μL Short Fragment Buffer (SFB). The wash step consisted of resuspension in SFB and subsequent magnetic separation of the washed beads. After the supernatant was removed, the pellet was resuspended in 13 μL 50 °C Elution Buffer. The elution mixture was incubated at 50 °C and 1000 rpm in a heater shaker for 10 min. After magnetic separation, 13 μL of the eluate was removed, and DNA content and purity were analyzed via Nanodrop. Sequencing libraries were created by combining 37.5 μL Sequencing Buffer, 25.5 μL Loading Beads, and 12 μL DNA Library.

2.5 Nanopore sequencing

In total, 37.5 μL of each library was used for sequencing on a Flongle flow Cell (R9.4.1). DNA contents were 48.6 ng for sample 1, containing fragments of all six plasmids, and 17.4 ng for sample 2, containing fragments of two plasmids, resembling two *IDH1* variants. Sequencing was concluded after 18 h, and a subset of sequences generated during the first 15 min was separated.

2.6 In silico analysis

To analyze the possible mutation sites bioinformatically, alignment references of the tumor marker sequences were created. In the reference sequences, the possible mutation site was deleted; therefore, alignment of each generated sequence produced an insertion mutation during variant calling. The bases recognized as insertions were used to distinguish wild-type and mutated sequences. Each generated sequence was aligned with all possible target references using minimap2 [51]. Subsequently, paftools. js was used for variant-calling [51] and custom scripts accumulated the numbers of wild-type and mutant reads in real time. To ensure the highest possible accuracy, matches between generated sequences and references were split into *mapped generated sequences* and *sequences with tumor marker information*. Because truncated sequences or erroneously sequenced DNA molecules can be aligned to a given reference but yield no tumor marker information, only generated sequences that can unambiguously be identified as a tumor marker variant were used for analysis.

2.7 Results and discussion

Creating a sequencing library from a mixture of six plasmids containing different tumor marker genes enabled us to identify each target with high accuracy, as seen in **Table 4**. These results were achieved after 18 h of sequencing with a library containing

Target	Inspected variation	Mapped generated Sequences	Sequences with tumor marker information	Correctly Identified Sequence variants [%]
IDH1_R132C	R132	128416	74002	99.57
IDH2_Wt	R172	196040	151329	97.59
pTERT_Wt	C228	48932	32613	99.45
	C250		34689	99.61
H3F3A_Wt	K27	39232	21899	99.11
	G34		22308	97.50
Hist1H3B_Wt	K27	45560	19458	99.66
BRAF_Wt	V600	52387	35354	98.36

Table 4.
A sequencing library was prepared from tumor marker DNA excised from synthetic plasmids. Equal amounts of plasmid were used for each target. ***Mapped sequences*** *were identified as a given marker sequence via minimap2, but only* ***sequences with tumor marker information*** *were able to be used for SNP calling. Shown is the cumulative output after 18 h of sequencing on a Flongle flow cell.*

Target	Inspected variation	Mapped generated Sequences	Sequences with tumor marker information	Correctly Identified Sequence variants [%]
IDH1_R132C	R132	914	604	99.50
IDH2_Wt	R172	894	757	98.94
pTERT_Wt	C228	348	257	100
	C250		269	100
H3F3A_Wt	K27	239	157	100
	G34		166	96.99
Hist1H3B_Wt	K27	292	143	100
BRAF_Wt	V600	246	185	100

Table 5.
A sequencing library was prepared from tumor marker DNA excised from synthetic plasmids. Equal amounts of plasmid were used for each target. ***Mapped sequences*** *were identified as a given marker sequence via minimap2, but only* ***sequences with tumor marker information*** *were able to be used for SNP calling. Shown is the cumulative output after 15 min of sequencing on a Flongle flow cell.*

48.6 ng of plasmid DNA. Overall yield of sequences was high with >39000 generated reads for all markers. *IDH1* and *IDH2* were outliers in this regard as they yielded 128416 and 196040 reads, respectively. As expected, not all generated sequences carried tumor marker information. The ratio of sequences with tumor marker information to all mapped sequences was 42–77%, depending on the given marker (**Table 4**). The ratio of correctly annotated tumor marker variants was >97% in all cases. We demonstrated the specificity of Cas9 cleavage by including a plasmid without target sequences as background in separate control experiments. No sequences derived from this control plasmid were generated during subsequent ONT sequencing.

Accuracy for simulated homozygosity and coverage of each marker after 18 h was very high. As time-to-result is a central parameter in clinical diagnostics, these results were examined regarding the coverage and accuracy after 15 min of sequencing time. This subset revealed a coverage for each marker of >200x after 15 min. Ratios of sequences with tumor marker information out of all generated sequences were similar after 15 min as compared with 18 h. They ranged from 48 to 84%, depending on the observed marker. Accuracy regarding the identification of the tumor marker sequence was also comparably high in this data subset. Between 96.99 and 100% of sequences with tumor marker information were annotated correctly, as shown in **Table 5**.

As these results were achieved with one variant per marker, no conclusions regarding heterozygosity detection were possible. Most mutations are heterozygous. Therefore, a subsequent experiment was performed to assess the analysis accuracy when including a simulated heterozygous mutation. For this experiment, a mixture of equal amounts of pUC57::IDH1_Wt and pUC57::IDH1_R132H was used for Cas9 sequencing. In total, 17.4 ng of plasmid DNA contained in the library was loaded onto the Flongle flow cell. Results shown in **Table 6** were achieved after 18 h of sequencing, with a subset of sequences generated during the first 15 min of sequencing being evaluated separately. In total, 625 reads were generated after 15 min of which 490 (72%) yielded tumor marker information. After 18 h, 29072 reads were generated and 20875 (78%) yielded tumor marker information. It was found that both *IDH1* variants present in the digested plasmid mix were detected during sequencing. The expected

	Sequencing time 15 min	Sequencing time 18 h
Mapped generated sequences	625	29072
Sequences with tumor marker information	490	20875
Expected ratio (Wt/R132H/other IDH1) [%]	50/50/0	50/50/0
Achieved ratio (Wt/R132H/other IDH1) [%]	40.2/59.59/0.21	38.49/61.4/0.01

Table 6.
A sequencing library was prepared from tumor marker DNA excised from synthetic plasmids. Equal amounts of plasmid were used for each target. ***Mapped sequences*** *were identified as a given marker sequence via minimap2, but only* ***sequences with tumor marker information*** *could be used for SNP calling. Shown is the cumulative output after 15 min and 18 h of sequencing on a Flongle flow cell.*

ratio of these variants was 50% IDH1_Wt and 50% IDH1_R132H, but an approximated 40%/60% split was observed after 15 min and 18 h.

Although Cas9 selective sequencing focuses on the sequence of interest and thus may lead to faster and more accurate results as compared with a whole genome sequencing approach, there exist some challenges. In a direct sequencing approach, there is no amplification step involved when analyzing only a single mutated base or area. Thus, one complete genome delivers only one read of the desired area, which in case of carcinoma mutations is mainly haploidic. Therefore, a huge amount of highly pure, high molecular weight, genomic DNA must be prepared and used, which depending on the amount of sample and method of nucleic acid preparation may be a limitation. Typically, the usage of 1–10 μg of human genomic DNA for Cas9 digestion is suggested (nCATS, [50]), corresponding to 150.000–1.500.000 copies of a diploid (female) genome, ideally resulting in the same number of reads. However, one must consider inefficiencies in Cas9 digestion, nucleic acid purification, and library preparation together with possible off-target digestion effects. Further on, there exists an intrinsic error rate of each sequencing method used and in case of cancerous tissue, it may consist of a mixture of wild-type and mutated cells depending on tumor heterogeneity and the general quality of tissue sampling. This may result in a general low number of reads, possibly beyond the coverage needed to safely identify a mutation.

Despite these drawbacks, we believe an nCATS-based approach to intra-surgical determination of a molecular tumor marker panel is justified, as it allows for live detection of marker variants including epigenetic information [50]. Preamplification of the targets might alleviate the high input DNA requirements but removes the ability to determine epigenetic properties of the sequences. That would render the effective analysis of markers such as MGMT methylation status, a predictive biomarker for efficacy of chemotherapy [53], impossible. PCR-based approaches such as qPCR would be very sensitive, as even a few copies of target DNA can produce a positive signal [54], but primer sets that incorporate the putative mutation site would be necessary to distinguish between wild-type and mutant sequences. This is a drawback in comparison to the chosen nCATS approach as this only detects anticipated mutations. Immunodetection of possible auto-antibodies (e.g., with ELISA) has been reported to be prone to false positives [55] and even though nanopore sequencing itself is prone to sequencing errors, they are distributed across the sequence, which leads to high consensus accuracy [56]. Drawbacks are the low resolution of homopolymers, which are prone to sequencing errors with the current flow cell generations. Second-generation sequencing would allow for high sensitivity and accuracy and is well-established but delivers only short sequences. Due to its sequencing by synthesis approach, epigenetic information is lost

Diagnostic tool	Mode of action	Advantages	Challenges	Accuracy	Sensitivity
nCATS [50]	Excision of target sequences from genome and subsequent sequencing	Native sequencing—epigenetic profile accessible, real-time basecalling	Large input DNA requirements, possible off-target sequencing	Medium, off-target digestions possible [58]; Nanopore sequencing, high intrinsic error rate [56]	Medium
(Reverse transcriptase) real-time PCR [3, 4]	Detection by amplification	Quantitative sequencing, well-established method	Reliant on primer specificity, affected by mutations in binding regions of primer probes	Very high	Very high
ELISA [7, 59]	Immuno-detection	Can detect a wide range of antigens and antibodies	Prone to false positives [55]	Low–medium	Medium
2nd Gen. Sequencing [57, 60]	Sequencing by synthesis	Established methods, high read count	Short read sequencing, high number of reads needed for contig generation	Very high	Very high
3rd Gen. Sequencing [37, 61]	ONT: electrical current changes PacBio: Fluorescence detection	Long read sequencing, easy assembly of long contigs	High intrinsic error rate (single sequence), low homopolymer resolution	Medium	Medium

Table 7.
Comparison of current diagnostic tools with nCATS.

in this case as well [57]. A comparison between mode of action, advantages and disadvantages, accuracy and sensitivity of those diagnostic tools is shown in **Table 7** below.

3. Conclusion

Considering the number of reads generated in 15 min using Flongle flow cells, the process might be sped up with the use of MinION flow cells. This would increase cost but cut sequencing time in a trade-off to be considered on a case-by-case basis. We demonstrated that the defined plasmid sequences could be analyzed via the described workflow. As the described experiments represent a work in progress, the use of isolated gDNA with significantly lower target sequence density as a template must be demonstrated. Previous works used either amplification-independent whole-genome sequencing (WGS) or amplicon sequencing to assess brain tumor marker variants. The WGS alternative is significantly slower due to the excess of non-target sequences generated but yielded additional epigenetic information for the regions of interest [62]. Enrichment of target sequences via PCR yields more favorable library compositions but eliminates epigenetic information [62]. As shown here, our

CRISPR/Cas9-based approach to enrich native target sequences might be able to combine the advantages of both previous strategies.

The results of the simulated heterozygosity were ~10% off from the expected 50% distribution of IDH1 Wt/IDH1 R132H, as shown in **Table 6**, but the fact that negligible amounts of other IDH1 mutations and no other tumor marker sequences were found is promising toward applications in clinical environments.

Ultimately, the goal of such a workflow should be a time-to-result below the time required for a neurosurgical tumor resection via craniotomy. This way, neurosurgeons could make informed decisions about the extent of the ongoing surgery and initiate personalized therapeutic modalities based on clinically actionable prognostic biomarkers [63]. Provided the RNPs are prepared in advance, the workflow described using the ONT Cas9 Sequencing kit, including 15 min time required for sequencing, would take 1:45 h altogether. Assuming gDNA extraction and preparative dephosphorylation add another 30–45 min [64], the total time-to-results may be as low as 2.5 h. Considering the lower abundancy of target sequences in gDNA compared with synthetic plasmids, sequencing times would likely increase to generate the same coverage, which must be accounted for. But this could partly be mitigated by the usage of a MinION flow cell instead of a Flongle flow cell. The analyzed marker panel can be amended by adding or subtracting crRNAs to target different sequences. A prerequisite for this approach is the presence of PAM sites in the vicinity of the new target genes. Mutations in AT-rich regions might be hard to access via SpCas9 because of its PAM-site requirement of 5'-NGG-3'. For this reason, different Cas9 proteins might be suitable candidates for amended workflows, such as xCas9, ScCas9-SC++, or an engineered FnCas9 [17–19]. In summary, these proof-of-concept results suggest that Cas9-aided targeted sequencing can generate diagnostically relevant tumor marker information in a short period of time and therefore might be a feasible diagnostic method for intra-surgical tumor diagnostics.

Acknowledgements

The authors gratefully thank Prof. Dr. Bruno Moerschbacher (University of Münster) and Prof. Dr. Wolfgang Streit (University of Hamburg) for their in-depth discussions and scientific input. We would also like to thank Sabina Schaal for the project management and scientific input. We gratefully acknowledge the financial support of the Federal Ministry of Education and Research (Funding indicator: 13GW0347A).

Author details

Maximilian Evers[1], Björn Brändl[2], Franz-Josef Müller[3], Sönke Friedrichsen[1] and Stephan Kolkenbrock[1*]

1 Altona Diagnostics GmbH, Hamburg, Germany

2 Max-Planck-Institute of Molecular Genetics, Berlin, Germany

3 University Hospital Schleswig Holstein, Kiel, Germany

*Address all correspondence to: stephan.kolkenbrock@altona-diagnostics.com

References

[1] Improved RT-qPCR could transform diagnostics [Internet]. 2020. Available from: https://www.nature.com/articles/d42473-020-00424-1. [Accessed: 2022-07-07]

[2] el Jaddaoui I, Allali M, Raoui S, Sehli S, Habib N, Chaouni B, et al. A review on current diagnostic techniques for COVID-19. Expert Review of Molecular Diagnostics. 2021;**21**:141-160. DOI: 10.1080/14737159.2021.1886927

[3] Skrzypski M. Quantitative reverse transcriptase real-time polymerase chain reaction (qRT-PCR) in translational oncology: Lung cancer perspective. Lung Cancer. 2008;**59**:147-154. DOI: 10.1016/j.lungcan.2007.11.008

[4] Ståhlberg A, Zoric N, Åman P, Kubista M. Quantitative real-time PCR for cancer detection: The lymphoma case. Expert Review of Molecular Diagnostics. 2005;**5**:221-230. DOI: 10.1586/14737159.5.2.221

[5] Matthijs G, Souche E, Alders M, Corveleyn A, Eck S, Feenstra I, et al. Guidelines for diagnostic next-generation sequencing. European Journal of Human Genetics. 2016;**24**:2-5. DOI: 10.1038/ejhg.2015.226

[6] Grumaz S, Stevens P, Grumaz C, Decker SO, Weigand MA, Hofer S, et al. Next-generation sequencing diagnostics of bacteremia in septic patients. Genome Medicine. 2016;**8**:1-13. DOI: 10.1186/s13073-016-0326-8

[7] Tighe PJ, Ryder RR, Todd I, Fairclough LC. ELISA in the multiplex era: Potentials and pitfalls. PROTEOMICS—Clinical Applications. 2015;**9**:406-422. DOI: 10.1002/prca.201400130

[8] Koczula KM, Gallotta A. Lateral flow assays. Essays in Biochemistry. 2016;**60**:111-120. DOI: 10.1042/EBC20150012

[9] Mustafa MI, Makhawi AM. Sherlock and detectr: CRISPR-cas systems as potential rapid diagnostic tools for emerging infectious diseases. Journal of Clinical Microbiology. 2021;**59**:e00745-e00720. DOI: 10.1128/JCM.00745-20

[10] Zuo Z, Liu J. Structure and dynamics of Cas9 HNH domain catalytic state. Scientific Reports. 2017;**7**:1-13. DOI: 10.1038/s41598-017-17578-6

[11] Hsu PD, Lander ES, Zhang F. Development and applications of CRISPR-Cas9 for genome engineering. Cell. 2014;**157**:1262-1278

[12] Ma Y, Zhang L, Huang X. Genome modification by CRISPR/Cas9. The FEBS Journal. 2014;**281**:5186-5193. DOI: 10.1111/febs.13110

[13] Hsu PD, Scott DA, Weinstein JA, Ran FA, Konermann S, Agarwala V, et al. DNA targeting specificity of RNA-guided Cas9 nucleases. Nature Biotechnology. 2013;**31**:827-832. DOI: 10.1038/nbt.2647

[14] Mali P, Esvelt KM, Church GM. Cas9 as a versatile tool for engineering biology. Nature Methods. 2013;**10**:957-963. DOI: 10.1038/nmeth.2649

[15] Jiang F, Doudna JA. CRISPR–Cas9 structures and mechanisms. Annual Review of Biophysics. 2017;**46**:505-529. DOI: 10.1146/annurev-biophys-062215-010822

[16] Chen YC, Sheng J, Trang P, Liu F. Potential application of the CRISPR/

CAS9 system against herpesvirus infections. Viruses. 2018;**10**:291. DOI: 10.3390/v10060291

[17] Hu JH, Miller SM, Geurts MH, Tang W, Chen L, Sun N, et al. Evolved Cas9 variants with broad PAM compatibility and high DNA specificity. Nature. 2018;**556**:57-63. DOI: 10.1038/nature26155

[18] Chatterjee P, Jakimo N, Lee J, Amrani N, Rodríguez T, Koseki SRT, et al. An engineered ScCas9 with broad PAM range and high specificity and activity. Nature Biotechnology. 2020;**38**:1154-1158. DOI: 10.1038/s41587-020-0517-0

[19] Hirano H, Gootenberg JS, Horii T, Abudayyeh OO, Kimura M, Hsu PD, et al. Structure and engineering of Francisella novicida Cas9. Cell. 2016;**164**:950-961. DOI: 10.1016/j.cell.2016.01.039

[20] Pardee K, Green AA, Takahashi MK, Braff D, Lambert G, Lee JW, et al. Rapid, low-cost detection of Zika Virus using programmable biomolecular components. Cell. 2016;**165**:1255-1266. DOI: 10.1016/j.cell.2016.04.059

[21] Auslander S, Fussenegger M. Toehold gene switches make big footprints. Nature. 2014;**516**:333-334. DOI: 10.1038/516333a

[22] Zhou W, Hu L, Ying L, Zhao Z, Chu PK, Yu XF. A CRISPR–Cas9-triggered strand displacement amplification method for ultrasensitive DNA detection. Nature Communications. 2018;**9**:5012. DOI: 10.1038/s41467-018-07324-5

[23] Huang M, Zhou X, Wang H, Xing D. Clustered regularly interspaced short palindromic repeats/Cas9 triggered isothermal amplification for site-specific nucleic acid setection. Analytical Chemistry. 2018;**90**:2193-2200. DOI: 10.1021/acs.analchem.7b04542

[24] Wang X, Xiong E, Tian T, Cheng M, Lin W, Sun J, et al. CASLFA: CRISPR/Cas9-mediated lateral flow nucleic acid assay. ACS Nano. 2020;**14**:2497-2508. DOI: 10.1021/acsnano.0c00022

[25] Azhar M, Phutela R, Kumar M, Ansari AH, Rauthan R, Gulati S, et al. Rapid and accurate nucleobase detection using FnCas9 and its application in COVID-19 diagnosis. Biosensors & Bioelectronics. 2021;**183**:113207. DOI: 10.1016/j.bios.2021.113207

[26] Quan J, Langelier C, Kuchta A, Batson J, Teyssier N, Lyden A, et al. FLASH: A next-generation CRISPR diagnostic for multiplexed detection of antimicrobial resistance sequences. Nucleic Acids Research. 2019;**47**:e83. DOI: 10.1093/nar/gkz418

[27] Bray F, Laversanne M, Weiderpass E, Soerjomataram I. The ever-increasing importance of cancer as a leading cause of premature death worldwide. Cancer. 2021;**127**:3029-3030. DOI: 10.1002/cncr.33587

[28] Hajian R, Balderston S, Tran T, de Boer T, Etienne J, Sandhu M, et al. Detection of unamplified target genes via CRISPR–Cas9 immobilized on a graphene field-effect transistor. Nature Biomedical Engineering. 2019;**3**:427-437

[29] Louis DN, Perry A, Wesseling P, Brat DJ, Cree IA, Figarella-Branger D, et al. The 2021 WHO Classification of Tumors of the Central Nervous System: A summary. Neuro-Oncology. 2021;**23**:1231-1251. DOI: 10.1093/neuonc/noab106

[30] Yang H, Ye D, Guan KL, Xiong Y. IDH1 and IDH2 mutations in tumorigenesis: Mechanistic insights

and clinical perspectives. Clinical Cancer Research. 2012;**18**:5562-5571. DOI: 10.1158/1078-0432.ccr-12-1773

[31] Powter B, Jeffreys SA, Sareen H, Cooper A, Brungs D, Po J, et al. Human TERT promoter mutations as a prognostic biomarker in glioma. Journal of Cancer Research and Clinical Oncology. 2021;**147**:1007-1017. DOI: 10.1007/s00432-021-03536-3

[32] Buczkowicz P, Hoeman C, Rakopoulos P, Pajovic S, Letourneau L, Dzamba M, et al. Genomic analysis of diffuse intrinsic pontine gliomas identifies three molecular subgroups and recurrent activating ACVR1 mutations. Nature Genetics. 2014;**46**:451-456. DOI: 10.1038/ng.2936

[33] Wu G, Broniscer A, McEachron TA, Lu C, Paugh BS, Becksfort J, et al. Somatic histone H3 alterations in paediatric diffuse intrinsic pontine gliomas and non-brainstem glioblastomas. Nature Genetics. 2012;**44**:251-253. DOI: 10.1038/ng.1102

[34] Pratilas CA, Xing F, Solit DB. Targeting oncogenic braf in human cancer. Current Topics in Microbiology and Immunology. 2012;**355**:83. DOI: 10.1007/82_2011_162

[35] Segerman B. The most frequently used sequencing technologies and assembly methods in different time segments of the bacterial surveillance and Refseq genome databases. Frontiers in Cellular and Infection Microbiology. 2020;**10**:527102

[36] Allali I, Arnold JW, Roach J, Cadenas MB, Butz N, Hassan HM, et al. A comparison of sequencing platforms and bioinformatics pipelines for compositional analysis of the gut microbiome. BMC Microbiology. 2017;**17**:194. DOI: 10.1186/s12866-017-1101-8

[37] Wang Y, Zhao Y, Bollas A, Wang Y, Au KF. Nanopore sequencing technology, bioinformatics and applications. Nature Biotechnology. 2021;**21**:1348

[38] Nicholls SM, Quick JC, Tang S, Loman NJ. Ultra-deep, long-read nanopore sequencing of mock microbial community standards. Gigascience. 2019;**8**:giz043

[39] Ardui S, Ameur A, Vermeesch JR, Hestand MS. Single molecule real-time (SMRT) sequencing comes of age: Applications and utilities for medical diagnostics. Nucleic Acids Research. 2018;**46**:2159-2168. DOI: 10.1093/nar/gky066

[40] Chen Z, He X. Application of third-generation sequencing in cancer research. Medical Review. 2021;**1**:150-171. DOI: 10.1515/mr-2021-0013

[41] Gouil Q, Keniry A. Latest techniques to study DNA methylation. Essays in Biochemistry. 2019;**63**:639-648

[42] Usui G, Matsusaka K, Mano Y, Urabe M, Funata S, Fukayama M, et al. DNA methylation and genetic aberrations in gastric cancer. Digestion. 2021;**102**:25-32. DOI: 10.1159/000511243

[43] Cao J, Yan Q. Cancer epigenetics, tumor immunity, and immunotherapy. Trends in Cancer. 2020;**6**:580-592. DOI: 10.1016/j.trecan.2020.02.003

[44] Villanueva L, Álvarez-Errico D, Esteller M. The contribution of epigenetics to cancer immunotherapy. Trends in Immunology. 2020;**11**:676 691. DOI: 10.1016/j.it.2020.06.002

[45] Li S, Tollefsbol TO. DNA methylation methods: Global DNA methylation and methylomic analyses. Methods. 2021;**187**:28-43. DOI: 10.1016/j.ymeth.2020.10.002

[46] Thudi M, Li Y, Jackson SA, May GD, Varshney RK. Current state-of-art of sequencing technologies for plant genomics research. Briefings in Functional Genomics. 2012;**1**:3-11. DOI: 10.1093/bfgp/elr045

[47] Amarasinghe SL, Su S, Dong X, Zappia L, Ritchie ME, Gouil Q. Opportunities and challenges in long-read sequencing data analysis. Genome Biology. 2020;**21**:30. DOI: 10.1186/s13059-020-1935-5

[48] Buytaers FE, Saltykova A, Denayer S, Verhaegen B, Vanneste K, Roosens NHC, et al. Towards real-time and affordable strain-level metagenomics-based foodborne outbreak investigations using oxford nanopore sequencing technologies. Frontiers in Microbiology. 2021;**21**:738284. DOI: 10.3389/fmicb.2021.738284

[49] Ebbert MTW, Farrugia SL, Sens JP, Jansen-West K, Gendron TF, Prudencio M, et al. Long-read sequencing across the C9orf72 "GGGGCC" repeat expansion: Implications for clinical use and genetic discovery efforts in human disease. Molecular Neurodegeneration. 2018;**13**:46. DOI: 10.1186/s13024-018-0274-4

[50] Gilpatrick T, Lee I, Graham JE, Raimondeau E, Bowen R, Heron A, et al. Targeted nanopore sequencing with Cas9-guided adapter ligation. Nature Biotechnology. 2020;**38**:433-438. DOI: 10.1038/s41587-020-0407-5

[51] Li H. Minimap2: Pairwise alignment for nucleotide sequences. Bioinformatics. 2018;**34**:3094-3100. DOI: 10.1093/bioinformatics/bty191

[52] Cone EB, Marchese M, Paciotti M, Nguyen DD, Nabi J, Cole AP, et al. Assessment of time-to-treatment initiation and survival in a cohort of patients with common cancers. JAMA Network Open. 2020;**3**:e2030072. DOI: 10.1001/jamanetworkopen.2020.30072

[53] Gerson SL. MGMT: Its role in cancer aetiology and cancer therapeutics. Nature Reviews. Cancer. 2004;**4**:296-307. DOI: 10.1038/nrc1319

[54] Fey A, Eichler S, Flavier S, Christen R, Höfle MG, Guzmán CA. Establishment of a real-time PCR-based approach for accurate quantification of bacterial RNA targets in water, using Salmonella as a model organism. Applied and Environmental Microbiology. 2004;**70**:3618-3623. DOI: 10.1128/aem.70.6.3618-3623.2004

[55] Garcia HH, Castillo Y, Gonzales I, Bustos JA, Saavedra H, Jacob L, et al. Low sensitivity and frequent cross-reactions in commercially available antibody-detection ELISA assays for Taenia solium cysticercosis. Tropical Medicine & International Health. 2018;**23**:101-105. DOI: 10.1111/tmi.13010

[56] Rang FJ, Kloosterman WP, de Ridder J. From squiggle to basepair: Computational approaches for improving nanopore sequencing read accuracy. Genome Biology. 2018;**19**:1-11. DOI: 10.1186/s13059-018-1462-9

[57] Grada A, Weinbrecht K. Next-generation sequencing: Methodology and application. The Journal of Investigative Dermatology. 2013;**133**:1-4. DOI: 10.1038/jid.2013.248

[58] Lin Y, Cradick TJ, Brown MT, Deshmukh H, Ranjan P, Sarode N, et al. CRISPR/Cas9 systems have off-target activity with insertions or deletions between target DNA and guide RNA sequences. Nucleic Acids Research. 2014;**42**:7473

[59] Tiscione NB. The validation of ELISA screening according to SWGTOX

recommendations. Journal of Analytical Toxicology. 2018;**42**:e33-e34

[60] Luo C, Tsementzi D, Kyrpides N, Read T, Konstantinidis KT. Direct comparisons of Illumina vs. Roche 454 sequencing technologies on the same microbial community DNA sample. PLoS ONE. 2012;7:2

[61] Roberts RJ, Carneiro MO, Schatz MC. The advantages of SMRT sequencing. Genome Biology. 2013;**14**:1-4

[62] Euskirchen P, Bielle F, Labreche K, Kloosterman WP, Rosenberg S, Daniau M, et al. Same-day genomic and epigenomic diagnosis of brain tumors using real-time nanopore sequencing. Acta Neuropathologica. 2017;**134**:691-703. DOI: 10.1007/s00401-017-1743-5

[63] Weerasinghe RK, Meng R, Dowdell AK, Bapat B, Vita A, Schroeder B, et al. Identification of clinically actionable biomarkers via routine comprehensive genomic profiling across a large community health system. Journal of Clinical Oncology. 2022;**40**:e15035. DOI: 10.1200/JCO.2022.40.16_suppl.e15035

[64] Seufi AEM, Galal FH. View of fast DNA purification methods: Comparative study. WAS Science Nature. 2020;**3**. Available from: http://worldascience.org/journals/index.php/wassn/article/view/9